GENETIC AND ENVIRONMENTAL MANIPULATION OF HORTICULTURAL CROPS

GENETIC AND ENVIRONMENTAL MANIPULATION OF HORTICULTURAL CROPS

Edited by

K.E. Cockshull, D. Gray, G.B. Seymour and B. Thomas

Horticulture Research International
Wellesbourne, UK

CABI *Publishing*

CABI *Publishing* – a division of CAB INTERNATIONAL

CABI *Publishing*
CAB INTERNATIONAL
Wallingford
Oxon OX10 8DE
UK

Tel: +44 (0)1491 832111
Fax: +44 (0)1491 833508
Email: cabi@cabi.org

CABI *Publishing*
10 E. 40th Street,
Suite 3203
New York, NY 10016
USA

Tel: +1 212 481 7018
Fax: +1 212 686 7993
Email: cabi-nao@cabi.org

© CAB INTERNATIONAL 1998. All rights reserved. No part of this publication may be reproduced in any form or by any means, electronically, mechanically, by photocopying, recording or otherwise, without the prior permission of the copyright owners.

A catalogue record for this book is available from the British Library, London, UK.

Library of Congress Cataloging-in-Publication Data
Genetic and environmental manipulation of horticultural crops / edited by K.E. Cockshull . . . [et al.].
    p. cm.
   Papers from the first Horticulture Research International conference, held at Wellesbourne in October 1997.
   Includes bibliographical references (p.) and index.
   ISBN 0-85199-281-1 (alk. paper)
   1. Horticultural crops--Congresses. 2. Horticultural crops--Genetics--Congresses. 3. Horticultural crops--Ecophysiology--Congresses. I. Cockshull, K. E. II. Horticulture Research International (Great Britain). Conference (1st : 1997 : Wellesbourne, UK)
SB317.53.G45   1998
635'.04233--dc21
                                                                                         98-25881
                                                                                           CIP

ISBN 0 85199 281 1

Typeset by York House Typographic Ltd, London
Printed and bound in the UK by Biddles, Guildford and King's Lynn

Contents

Contributors		vii
Preface		xi
1	Genetic Approaches to Manipulation of Fruit Development and Quality in Tomato *J.J. Giovannoni, P. Kannan, S. Lee and H.C. Yen*	1
2	Improving Tomato Fruit Quality by Cultivation *L.C. Ho*	17
3	GCRI/Bewley Lecture: Applications of Molecular Biology and Genetic Manipulation to Understand and Improve Quality of Fruits and Vegetables *D. Grierson*	31
4	Gene Expression in Ripening Bananas *R. Drury, C.R. Bird and G.B. Seymour*	41
5	Genes for Fruit Quality in Strawberry *K. Manning*	51
6	The Tomato Ethylene Receptor Gene Family: It's Not Easy Being a Plant *D. Tieman and H. Klee*	63
7	Environmental Requirements as Determined by Rooting Potential in Leafy Cuttings *R.S. Harrison-Murray and B.H. Howard*	75

8 **The Use of Mutants and Molecular Biology to Understand Competence for Root Formation** 95
   W.P. Hackett, S.T. Lund, A.G. Smith and N.E. Olszewski

9 **Physiological Analysis of the Floral Transition** 103
   G. Bernier, L. Corbesier, C. Périlleux, A. Havelange and P. Lejeune

10 **Genetic and Environmental Control of Flowering in Strawberry** 111
   N.H. Battey, P. Le Mière, A. Tehranifar, C. Cekic, S. Taylor, K.J. Shrives, P. Hadley, A.J. Greenland, J. Darby and M.J. Wilkinson

11 **Manipulating the Photoperiodic Control of Plant Reproduction** 133
   S.D. Jackson and B. Thomas

12 **Regulation of Abscisic Acid and Water Stress Response Genes** 143
   P.K. Busk, M. Figueras, A.C. Jessop, A. Goday and M. Pagès

13 **Manipulation of Growth of Horticultural Crops Under Environmental Stress** 157
   W.J. Davies, D.S. Thompson and J.E. Taylor

14 **Engineering Phytochrome Genes to Improve Crop Performance** 175
   H. Smith

15 **Regulation of Stem Extension by Temperature** 191
   F.A. Langton

16 **Modification of Plant Morphology by Genetic Manipulation of Gibberellin Biosynthesis** 205
   P. Hedden, J.P. Coles, A.L. Phillips, S.G. Thomas, D.A. Ward, I.S. Curtis, J.B. Power, K.C. Lowe and M.R. Davey

Index 219

Contributors

N.H. Battey, Soft Fruit Technology Group, School of Plant Sciences, The University of Reading, Whiteknights, Reading RG6 6AS, UK.

G. Bernier, Laboratoire de Physiologie Végétale, Université de Liège, Sart Tilman, B4000 Liège, Belgium.

C.R. Bird, Zeneca Plant Science, Jealott's Hill Research Station, Bracknell, Berks RG12 6EY, UK.

P.K. Busk, Departament de Genética Molecular, C.I.D. (C.S.I.C.), Jordi Girona 18-26, 08034 Barcelona, Spain.

C. Cekic, School of Plant Sciences, University of Reading, Whiteknights, Reading RG6 6AS, UK.

K.E. Cockshull, Horticulture Research International, Wellesbourne, Warwick CV35 9EF, UK.

J.P. Coles, IACR-Long Ashton Research Station, Department of Agricultural Sciences, University of Bristol, Long Ashton, Bristol BS41 9AF, UK.

L. Corbesier, Laboratoire de Physiologie Végétale, Université de Liège, Sart Tilman, B4000 Liège, Belgium.

I.S. Curtis, Plant Science Division, School of Biological Sciences, University of Nottingham, University Park, Nottingham NG7 2RD, UK.

J. Darby, Darby Brothers Farms Ltd, Methwold Hythe, Norfolk IP26 4PU, UK.

M.R. Davey, Plant Science Division, School of Biological Sciences, University of Nottingham, University Park, Nottingham NG7 2RD, UK.

W.J. Davies, Department of Biological Sciences, I.E.N.S., Lancaster University, Bailrigg, Lancaster LA1 4YQ, UK.

R. Drury (née Medina-Suárez), Horticulture Research International, Wellesbourne, Warwick CV35 9EF, UK. (Present address: ICRF Molecular Oncology Unit, Department of Cancer Medicine, ICSM at Hammersmith, DuCane Road, London W12 0NN, UK.)

M. Figueras, Departament de Genética Molecular, C.I.D. (C.S.I.C.), Jordi Girona 18-26, 08034 Barcelona, Spain.

J.J. Giovannoni, Department of Horticultural Sciences and Crop Biotechnology Center, Texas A&M University, College Station, TX 77843-2133, USA.

A. Goday, Departament de Genética Molecular, C.I.D. (C.S.I.C.), Jordi Girona 18–26, 08034 Barcelona, Spain.

D. Gray, Horticulture Research International, Wellesbourne, Warwick CV35 9EF, UK.

A.J. Greenland, ZENECA Agrochemicals, Jealott's Hill Research Station, Bracknell RG42 6EY, UK.

D. Grierson, Plant Science Division, School of Biological Sciences, University of Nottingham, Sutton Bonington Campus, Loughborough, Leics LE12 5RD, UK.

W.P. Hackett, Departments of Horticultural Science and Plant Biology, University of Minnesota, St Paul, MN 55108, USA.

P. Hadley, School of Plant Sciences, University of Reading, Whiteknights, Reading RG6 6AS, UK.

R.S. Harrison-Murray, Horticulture Research International, East Malling, Maidstone, Kent ME19 6BJ, UK.

A. Havelange, Laboratoire de Physiologie Végétale, Université de Liège, Sart Tilman, B4000 Liège, Belgium.

P. Hedden, IACR-Long Ashton Research Station, Department of Agricultural Sciences, University of Bristol, Long Ashton, Bristol BS41 9AF, UK.

L.C. Ho, Horticulture Research International, Wellesbourne, Warwick CV35 9EF, UK.

B.H. Howard, Horticulture Research International, East Malling, Maidstone, Kent ME19 6BJ, UK.

S.D. Jackson, Horticulture Research International, Wellesbourne, Warwick CV35 9EF, UK.

A.C. Jessop, Departament de Genética Molecular, C.I.D. (C.S.I.C.), Jordi Girona 18–26, 08034 Barcelona, Spain.

P. Kannan, Department of Horticultural Sciences and Crop Biotechnology Center, Texas A&M University, College Station, TX 77843-2133, USA.

H. Klee, Department of Horticultural Sciences, University of Florida, 1143 Fifield Hall, Gainesville, FL 32611, USA.

F.A. Langton, Horticulture Research International, Wellesbourne, Warwick CV35 9EF, UK.

P. Le Mière, School of Plant Sciences, University of Reading, Whiteknights, Reading RG6 6AS, UK.

S. Lee, Department of Horticultural Sciences and Crop Biotechnology Center, Texas A&M University, College Station, TX 77843-2133, USA.

P. Lejeune, Laboratoire de Physiologie Végétale, Université de Liège, Sart Tilman, B4000 Liège, Belgium.

K.C. Lowe, Plant Science Division, School of Biological Sciences, University of Nottingham, University Park, Nottingham NG7 2RD, UK.

S.T. Lund, Departments of Horticultural Science and Plant Biology, University of Minnesota, St Paul, MN 55108, USA.

K. Manning, Horticulture Research International, Wellesbourne, Warwick CV35 9EF, UK.

N.E. Olszewski, Departments of Horticultural Science and Plant Biology, University of Minnesota, St Paul, MN 55108, USA.

M. Pagès, Departament de Genética Molecular, C.I.D. (C.S.I.C.), Jordi Girona 18–26, 08034 Barcelona, Spain.

C. Périlleux, Laboratoire de Physiologie Végétale, Université de Liège, Sart Tilman, B4000 Liège, Belgium.

A.L. Phillips, IACR-Long Ashton Research Station, Department of Agricultural Sciences, University of Bristol, Long Ashton, Bristol BS41 9AF, UK.

J.B. Power, Plant Science Division, School of Biological Sciences, University of Nottingham, University Park, Nottingham NG7 2RD, UK.

G.B. Seymour, Horticulture Research International, Wellesbourne, Warwick CV35 9EF, UK.

K.J. Shrives, Farms Advisory Services Team (FAST), Faversham, Kent ME13 0LN, UK.

A.G. Smith, Departments of Horticultural Science and Plant Biology, University of Minnesota, St Paul, MN 55108, USA.

H. Smith, Department of Biology, University of Leicester, Leicester LE1 7RH, UK.

J.E. Taylor, Department of Biological Sciences, I.E.N.S., Lancaster University, Bailrigg, Lancaster LA1 4YQ, UK.

S. Taylor, School of Plant Sciences, University of Reading, Whiteknights, Reading RG6 6AS, UK.

A. Tehranifar, School of Plant Sciences, University of Reading, Whiteknights, Reading RG6 6AS, UK.

D. Tieman, Department of Horticultural Sciences, University of Florida, 1143 Fifield Hall, Gainesville, FL 32611 USA.

B. Thomas, Horticulture Research International, Wellesbourne, Warwick CV35 9EF, UK.

S.G. Thomas, IACR-Long Ashton Research Station, Department of Agricultural Sciences, University of Bristol, Long Ashton, Bristol BS41 9AF, UK.

D.S. Thompson, Department of Biological Sciences, I.E.N.S., Lancaster University, Bailrigg, Lancaster LA1 4YQ, UK.

D.A. Ward, IACR-Long Ashton Research Station, Department of Agricultural Sciences, University of Bristol, Long Ashton, Bristol BS41 9AF, UK.

M.J. Wilkinson, School of Plant Sciences, University of Reading, Whiteknights, Reading RG6 6AS, UK.

H.C. Yen, Department of Horticultural Sciences and Crop Biotechnology Center, Texas A&M University, College Station, TX 77843-2133, USA.

Preface

Advances in science and technology are continually creating exciting opportunities for novel and improved products and systems of plant production. Foremost amongst these in recent years have been advances in our understanding of how genes control plant growth and development. Much attention has been paid to how these advances might be used in agriculture, but relatively little attention has been paid to their possible exploitation in horticulture. This is surprising, since horticulture not only operates on a smaller physical scale than agriculture but it also uses a much larger number of different species and it has a more plant-centred tradition that involves the handling and manipulation of plants. Most of these features have the potential greatly to enhance what genetic manipulation alone can achieve and to add value to a commercial crop.

The protected cultivation of horticultural crops, especially in glasshouses, also requires a considerable measure of environmental control as well as a detailed knowledge of the responses of the crops to their environment. Although it is a truism that plant responses are the product of interactions between the plant's genotype and its environment, it is essential that we understand the environmental responses of new genotypes if we are to obtain maximum value from them. For all of the above reasons, therefore, it might seem that horticultural crop production was particularly well suited to making use of scientific and technical discoveries involving the genetic and environmental manipulation of crop plants.

Horticulture Research International (HRI) was established in 1990 to create and take advantage of new opportunities for horticultural research and development in the UK. It was formed by taking the three horticultural research institutes of the then Agricultural and Food Research Council, i.e. the former East Malling Research Station, Glasshouse Crops Research Institute (GCRI) and National Vegetable Research Station, which had come together with the Department of Hop Research at Wye College to form the Institute of Horticultural

Research, and then amalgamating them with the three remaining Experimental Horticulture Stations, i.e. Efford, Kirton and Stockbridge House of the Agricultural Development and Advisory Service. This process required considerable restructuring over a number of years, and the Chief Executive of HRI (Professor C.C. Payne) decided that one way to celebrate the successful establishment of this new body would be to hold an International Conference at HRI's strategic science centre in Wellesbourne, Warwickshire.

The organizers of the conference promptly proposed that it should aim to highlight opportunities where scientific advances in our understanding of key biological processes could lead to commercial applications involving the genetic or environmental manipulation of horticultural crops. Furthermore, they proposed that the conference should also aim to present recent examples that illustrated the progression from advances in basic science through to practical advances that benefited producers and consumers alike. Fortunately, Professor Payne not only agreed to the concept but gave it his wholehearted support and encouraged us to invite prominent authorities in relevant fields of research not only from the UK but also from the USA and from Europe. We then had the double good fortune that the Trustees of the GCRI Trust agreed to the prestigious GCRI/Bewley Lecture being part of the conference, and Professor Donald Grierson of the University of Nottingham accepted the Trustee's invitation to deliver the GCRI/Bewley Lecture entitled 'Applications of molecular biology and genetic manipulation to understand and improve quality of fruits and vegetables'.

This book contains much of the proceedings of the HRI Conference which was successfully held on 30 and 31 October 1997. The Editors are most grateful to Mrs Ann Beeny who was the Secretary to the Conference, as well as our link to the authors of these contributions, and who undertook the awesome task of collating the manuscripts into a common format.

Ken Cockshull

1 Genetic Approaches to Manipulation of Fruit Development and Quality in Tomato

J.J. GIOVANNONI, P. KANNAN, S. LEE AND H.C. YEN

Department of Horticultural Sciences and Crop Biotechnology Center, Texas A&M University, College Station, TX 77843-2133, USA

Introduction

Fruit ripening represents a biological process unique to plant species in which developmental and hormonal signalling systems orchestrate a variety of biochemical and physiological changes which in summation result in the 'ripe' stage of fruit maturation. In so-called 'climacteric' fruits such as tomato, cucurbits, banana, apple and many others, the initiation of ripening is characterized by a dramatic increase in respiration and biosynthesis of the gaseous hormone ethylene (Abeles *et al.*, 1992). Inhibition of ethylene biosynthesis or ethylene perception via exogenous application of inhibitors, or endogenous expression of transgenes, has been shown to have profound inhibitory effects on ethylene-mediated plant processes including climacteric fruit ripening (reviewed in Mattoo and Suttle, 1991; Abeles *et al.*, 1992; Theologis, 1992; Giovannoni, 1993; Hobson and Grierson, 1993; Gray *et al.*, 1994; Ecker, 1995). As a result, the majority of scientific effort devoted to ripening research has been in the areas of ethylene biosynthesis, ethylene responses and, more recently, ethylene perception and signal transduction (Wilkinson *et al.*, 1995; Yen *et al.*, 1995). Nevertheless, careful examination of the physiological, biochemical, genetic and molecular data collected in recent decades (particularly from analysis of ripening tomato fruit) suggests a significant developmental component of fruit ripening which interacts with and modulates ethylene biosynthesis and signalling during ripening, and has been largely overlooked (Theologis *et al.*, 1993; Wilkinson *et al.*, 1995). We and others have taken a multifaceted molecular and genetic approach toward elucidation of the genetic determinants which underlie this complex regulatory process.

Fruit ripening is a complex quantitative trait

The analysis of single gene mutations and transgenic plants repressed for particular ripening-related genes clearly demonstrates that single loci can have profound effects on fruit ripening and quality (reviewed in Giovannoni, 1993; Gray et al., 1994). However, the fact that numerous genes have been shown to influence fruit development clearly defines ripening and fruit quality as quantitative traits. In 1988 and 1991, Paterson et al. described the first detailed analysis of fruit quality qualitative trait loci (QTLs) through definition of loci that influenced fruit mass, solids and acidity. Identification of DNA markers linked to QTLs with a substantial impact on trait phenotype represent molecular tools both for modification of quality via molecular breeding, and for targeting and eventual isolation of genes likely to have the greatest impact on the said characteristics.

More recently, Alpert et al. (1995) demonstrated that nearly isogenic lines distinguished by a single QTL influencing fruit mass could be generated, thus resulting in the practical conversion of a major QTL into a single gene trait. The implications of this finding are profound, as this result indicates that molecular markers can be used to generate lines for physiological analysis of effects of single QTLs, and the resulting lines may also permit isolation of QTLs via map-based cloning. Advanced back-cross lines resulting from crosses between the cultivated tomato (*Lycopersicon esculentum*) and its wild relative *L. pimpinellifolium* subsequently have been created for analysis of virtually any QTL, including those related to fruit ripening and quality, for which there is variation between the population parents (Tanksley and Nelson, 1996; Tanksley et al., 1996). Availability of such lines will probably have a significant impact on the ability to assess the number and relative influence of loci contributing to fruit ripening and quality in the near future, and will eventually lead to the cloning and characterization of the corresponding genes.

Ripening molecular biology and ethylene signalling in tomato

The critical role of ethylene in coordinating climacteric ripening at the molecular level was first observed via analysis of ethylene-inducible ripening-related gene expression (reviewed in Giovannoni, 1993). Ripening-related genes have been isolated via differential gene expression patterns (Slater et al., 1985; Lincoln et al., 1987) and biochemical function (DellaPenna et al., 1986; Sheehy et al., 1987; Biggs and Handa, 1989; Harriman et al., 1991; Oeller et al., 1991; Yelle et al., 1991). Promoter analysis of ripening genes has been performed via examination of promoter–reporter construct activities in transient assays and transgenic plants. The result has been the identification of *cis*-acting promoter elements and *trans*-acting factors which are responsible for both ethylene- and non-ethylene-regulated aspects of ripening (Deikman et al., 1992; Montgomery et al., 1993).

The *in vivo* functions of several ripening-related gene products including polygalacturonase (PG), pectin methylesterase (PME), 1-aminocyclopropane-1-carboxylate (ACC) synthase, ACC oxidase and phytoene synthase have been tested via antisense gene repression and/or mutant complementation in tomato. For example, the cell wall pectinase, PG, was shown to be necessary for ripening-related pectin depolymerization and pathogen susceptibility, though with little effect on fruit softening (Smith *et al.*, 1988; Giovannoni *et al.*, 1989; Kramer *et al.*, 1990). Inhibition of phytoene synthase resulted in reduced carotenoid biosynthesis and reduction in fruit and flower pigmentation (Fray and Grierson, 1993). Reduced ethylene evolution resulting in ripening inhibition occurred with antisense ACC synthase and ACC oxidase (Hamilton *et al.*, 1990; Oeller *et al.*, 1991).

Developmental cues coordinate the ethylene response during ripening

Analysis of both naturally occurring mutants and transgenic tomatoes inhibited in ethylene biosynthesis demonstrates that climacteric ripening responds to a combination of ethylene regulation and developmental control. Although antisense ACC synthase tomatoes which failed to produce ethylene did not ripen, several ripening-related genes were still expressed. This observation supports the presence of a developmental (non-ethylene-regulated) component of ripening. Indeed the gene encoding the rate-limiting activity in ethylene biosynthesis, ACC synthase, is itself initially induced during ripening by a developmental signal (Theologis *et al.*, 1993).

Gene expression in the non-ripening *rin* and *nor* tomato mutants is also impaired for most ripening-related genes examined (DellaPenna *et al.*, 1989; Picton *et al.*, 1993). In addition to demonstrating a lack of ethylene-inducible gene expression (due to the lack of climacteric ethylene), similar to that observed in transgenic ethylene-reduced fruit, the *rin* and *nor* mutants are also deficient in expression of genes whose regulation is at least partly developmental in nature such as PG, E8 and ACC synthase (DellaPenna *et al.*, 1989; Theologis *et al.*, 1993). This fact, in combination with the observations that: (i) *rin* and *nor* fruit ripening inhibition cannot be reversed with exogenous ethylene, and (ii) many ethylene-regulated genes can be induced in mutant fruit with the application of exogenous ethylene (demonstrating that *rin* and *nor* are ethylene responsive), strongly supports the hypothesis that *rin* and *nor* represent lesions influencing the developmentally regulated component of climacteric ripening. Alternatively, *rin* and *nor* may represent steps downstream of primary ethylene signalling as does the *Arabidopsis hookless* (*hls*) gene (Lehman *et al.*, 1996); however, the lack of ethylene biosynthesis and inhibited expression of developmentally influenced genes such as PG, E8 and ACC synthase argue in favour of the former hypothesis.

Genetic characterization of ethylene signal transduction in tomato

Lanahan *et al.* (1994) observed that the *Nr* mutation is manifested as a block in a host of ethylene responses including inhibition of the seedling triple response and incomplete fruit ripening. The *Nr* gene was cloned, sequenced and shown to have high homology to the *ETR1* and *ERS* genes of *Arabidopsis* (Wilkinson *et al.*, 1995). All three genes have significant homology to members of the 'two-component' class of protein kinases (Koshland, 1995). The reduction of ethylene-binding capacity in *Etr1* mutants (Bleeker *et al.*, 1988) and yeast expressing a mutant *ETR1* cDNA (Schaller and Bleeker, 1995), in combination with the fact that the *ETR1* gene product apparently is involved in protein phosphorylation, suggests that *ETR1* (and presumably *Nr* and *ERS*) is likely to encode an ethylene receptor which conveys signal transduction via modification of protein phosphorylation.

Nr gene expression is distinct from the constitutive expression of *ETR1* and *ERS* in that *Nr* expression is itself ethylene inducible, and the ethylene responsiveness of *Nr* additionally is dependent on the developmental stage of the fruit (Wilkinson *et al.*, 1995). This observation provides additional evidence for a significant role of non-ethylene developmental signals in coordinating climacteric fruit ripening, and demonstrates that such signals are important for the regulation of ethylene signalling.

The tomato *Epi* mutant is likely to represent an additional component of ethylene signal transduction (Fujino *et al.*, 1989). The *Epi* mutation originally was characterized as a semi-dominant, single locus mutation resulting in leaf epinasty, vertical growth, minimal branching and highly branched root structure. These effects are consistent with ethylene overproduction or constitutive ethylene signalling (Ecker, 1995). Although elevated ethylene biosynthesis has been reported in some tissues of the *Epi* mutant, treatment with inhibitors of ethylene biosynthesis or action had little effect on mutant phenotype, suggesting that *Epi* represents a lesion in ethylene signalling (Fujino *et al.*, 1989). The *Arabidopsis ctr1* mutant is also characterized by constitutive ethylene signal transduction, and the corresponding *CTR1* gene has been isolated and shown to have homology to the *Raf* family of protein kinases (Kieber *et al.*, 1993). It is possible that *Epi* may represent a tomato homologue of *CTR1*. Alternatively, *Epi* may represent a unique ethylene signal transduction component. As described below, much of our recent work has centred on molecular and genetic analysis of the *Epi* mutant and genetic attempts to address the question of whether or not the lesion resulting in the *Epi* mutation corresponds to a mutation in a *CTR1* homologue.

Genetic Mapping of Single Gene Mutations Influencing Tomato Fruit Ripening and Quality

As a first step toward isolation of several genes known via mutation to have significant impacts on fruit ripening and quality, we have developed large F2 mapping populations segregating for the *rin*, *nor*, *Nr* and *hp* loci. DNA markers described by Tanksley *et al.* (1992), random amplified polymorphic DNAs (RAPDs), and amplified fragment length polymorphism (AFLPs) were utilized to develop saturated restriction fragment length polymorphism (RFLP) maps around all four target loci, resulting in the accurate placement of *rin*, *nor*, *Nr* and *hp* on chromosomes 5, 10, 9 and 2, respectively. The details of these mapping studies have been described previously (Giovannoni *et al.*, 1995; Yen *et al.*, 1995, 1997) and are summarized in Fig. 1.1.

The mapping of the *Nr* locus to chromosome 9, combined with the

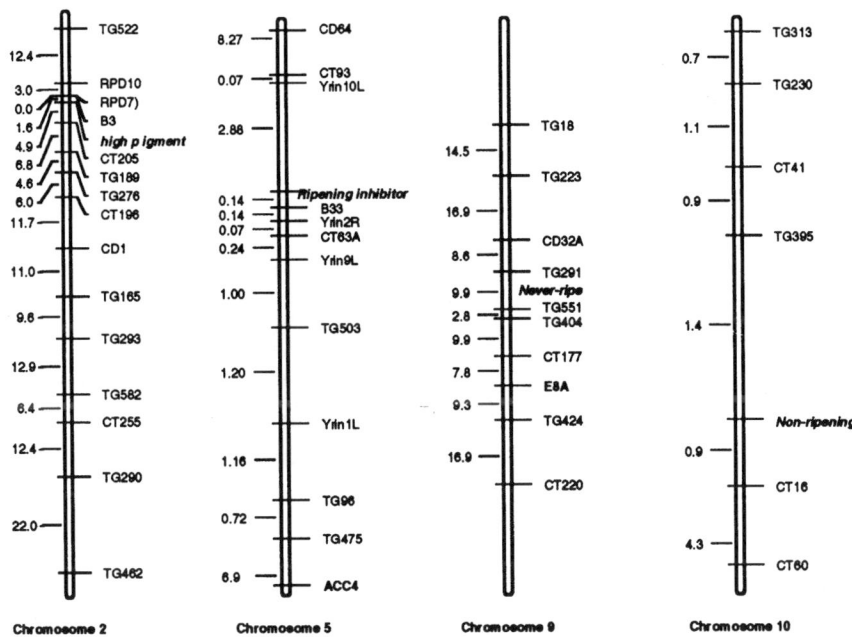

Fig. 1.1. Genetic maps of tomato fruit ripening loci. Numbers to the left of each chromosome correspond to genetic distances (in cM) between adjacent loci.

mapping of several *ETR1*-like loci in tomato, supported the hypothesis that *Nr* did in fact represent an ethylene receptor mutation, as was demonstrated ultimately by Wilkinson *et al.* (1995). Mutations at the *rin* and *nor* loci are phenotypically similar in that both essentially arrest ripening at the mature green stage when present in the homozygous state (Tigchelaar *et al.*, 1978). High-resolution mapping of both mutations has resulted in isolation of tomato yeast artificial chromosome (YAC) clones harbouring both loci (Giovannoni *et al.*, 1995). Recent efforts have resulted in the localization of the *rin* locus to a 95-kb region which currently is being reconstructed in a cosmid contig (V. Padmanabahn and J. Giovannoni, unpublished), and the isolation of 15 candidate cDNAs which are being tested for linkage to the *nor* locus (J. Vrebalov and J. Giovannoni, unpublished).

Mapping of the *hp* locus suggests tight linkage to the 45S ribosomal repeat region of chromosome 2, which may complicate map-based cloning of this locus. Nevertheless, phenotypic analysis of *hp* and normal fruit controls has not been impaired, and suggests that *hp* influences aspects of fruit quality not previously reported. In particular, *hp/hp* fruit accumulate significant levels of sucrose as compared with controls (which accumulate virtually no sucrose), and are elevated approximately 20-fold in accumulation of the antioxidant flavonoid quercetin (Fig. 1.2). In addition, we have shown that *hp* hypocotyls have an increased ratio of plastid to genomic DNA and increased plastid numbers (Yen *et al.*, 1997). Finally, the fact that *hp* represents a negative regulator of light signal

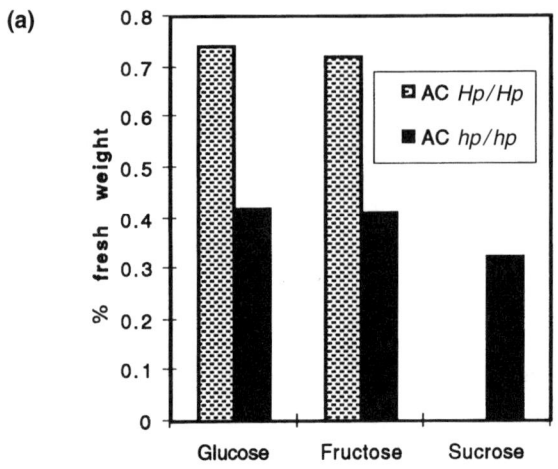

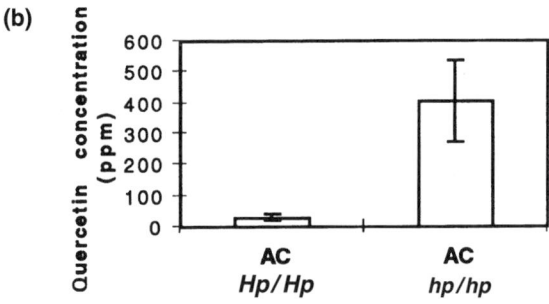

Fig. 1.2. Quality characteristics of normal and *hp* fruit. (a) Quercetin and (b) sugar concentrations in normal and nearly isogenic mutant fruit pericarp harvested 7 days post-breaker stage.

transduction suggests the possibility that this gene may correspond to light signal transduction loci described in *Arabidopsis*. This hypothesis is being tested via low-stringency RFLP mapping of cloned *Arabidopsis* light signalling genes relative to tomato DNA marker loci.

Genetic and Molecular Analysis of Ethylene Signal Transduction in Tomato

Isolation and characterization of a tomato *CTR1*-like gene: *TCTR1*

DNA gel-blot hybridization of the *Arabidopsis CTR1* full-length cDNA to tomato genomic DNA (at high stringency) indicates the presence of one to three related

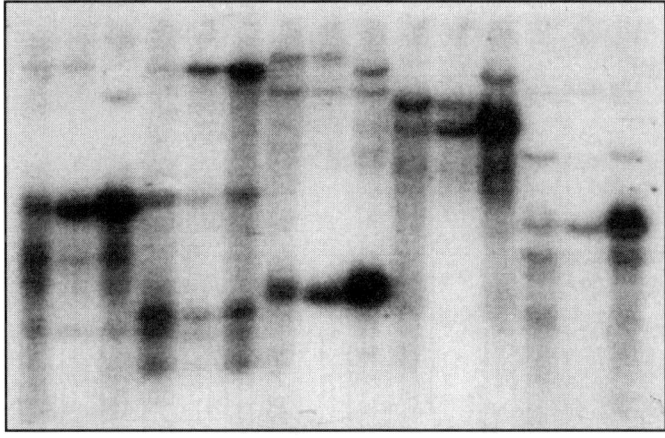

Fig. 1.3. DNA gel-blot of tomato genomic DNAs probed with the *TCTR1* cDNA. Genomic DNA was isolated from *L. esculentum*, *L. cheesmannii* and *L. pinnellii* leaves and digested with the indicated enzymes prior to electrophoresis and blotting.

sequences in the tomato genome (Fig. 1.3). In addition, one band appeared predominantly with each of the restriction enzymes tested, suggestive of one highly homologous locus. A total of 100,000 clones from a breaker stage tomato fruit cDNA library were screened using the *CTR1* cDNA as probe, and resulted in the isolation of a 1.3-kb cDNA which hybridizes to the *Arabidopsis CTR1* cDNA at high stringency. 5' and 3' rapid amplification of cDNA ends–polymerase chain reaction (RACE-PCR) were employed to isolate near full-length overlapping sequences which have been fused to form a single contiguous cDNA. The tomato cDNA has 70% DNA sequence identity and 62% amino acid identity with *CTR1*. We have tentatively named this cDNA *TCTR1* (P. Kannan and J. Giovannoni, unpublished).

Preliminary gene expression analysis indicates that the *TCTR1* transcript is approximately 3 kb in length, which is similar to that reported for *CTR1*. Of particular interest was the observation that *TCTR1* is induced during ripening, and by exogenous ethylene in mature green fruit (Fig. 1.4). *TCTR1* additionally is ethylene inducible in the *rin* and *nor* mutants, and expressed at lower levels in both ripening and ethylene-treated *Nr* fruit. This pattern of expression in fruit is similar to that observed for the *Nr* ethylene receptor (Wilkinson *et al.*, 1995) and distinct from the constitutive expression pattern reported for *CTR1* (Kieber *et al.*, 1993). Nevertheless, *TCTR1* apparently is constitutively expressed in total RNA derived from leaves and seedlings (S. Lee and J. Giovannoni, unpublished).

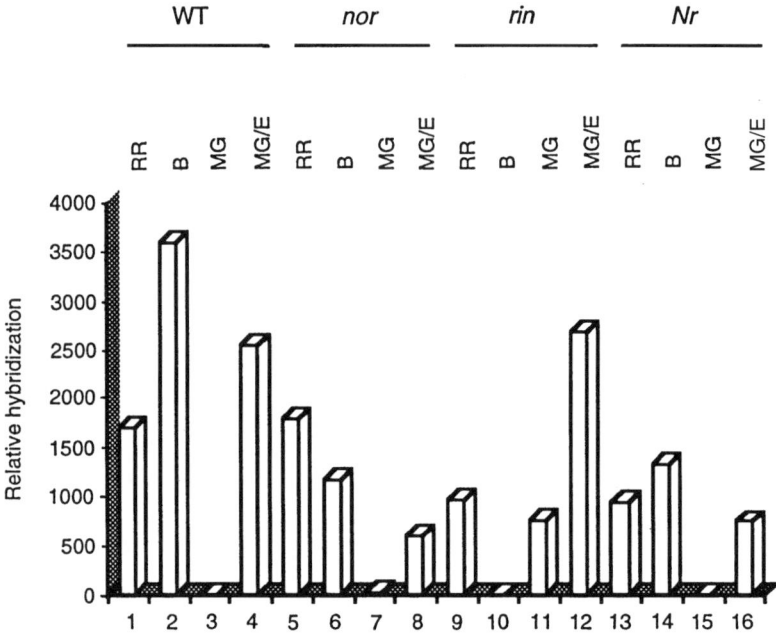

Fig. 1.4. Relative expression of *TCTR1* in normal and mutant tomato fruit. Red-ripe (RR), breaker (B), mature green (MG) and ethylene-treated (8 h with 10 p.p.m. ethylene) mature green (MG/E) fruit from normal and equivalent age mutant fruit were extracted for RNA and analysed via RNA gel-blot analysis. The resulting autoradiograph was scanned and relative intensities are shown. Units of hybridization are arbitrary and have no intrinsic meaning.

Together, these results suggest that *TCTR1* is constitutively expressed in some tissues, or during particular stages of development, and modulated by developmental and/or hormonal signals in other tissues or developmental stages. Additional regulation of putative (*TCTR1*) and known (*NR*) ethylene signal transduction components during tomato fruit development suggests that ethylene signal transduction is more tightly regulated in tissues whose proper development is largely dependent on effects resulting from ethylene hormone action. In addition, ethylene inducibility of at least two components of ethylene signal transduction, in addition to genes encoding rate-limiting steps in ethylene biosynthesis, provides a molecular explanation for the well-documented observations of autocatalytic ethylene biosynthesis and enhanced ethylene sensitivity following initial hormone exposure (Fig. 1.5).

Molecular, genetic and physiological characterization of the *Epi* mutant: a putative constitutive ethylene response mutation in tomato

Tomato plants homozygous for the mutant *Epi* allele are characterized by vertical growth, minimal lateral branching and leaf epinasty (Ursin, 1987). In addition, *Epi* seedlings demonstrate a constitutive triple-response phenotype as

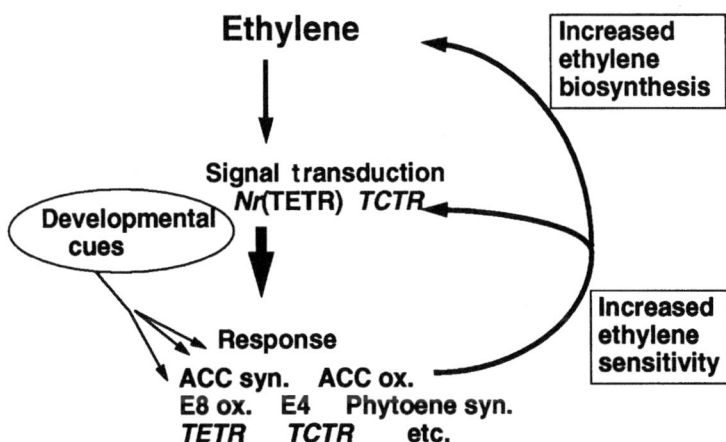

Fig. 1.5. Autocatalytic ethylene biosynthesis and enhanced hormone-induced sensitivity can be explained via the ethylene-inducible nature of ethylene biosynthetic and signal transduction genes.

was previously shown by Fujino et al. (1989). While *Epi* tissues are known to overproduce ethylene, treatment of seedlings with inhibitors of ethylene action or biosynthesis did not result in reversion of the mutant phenotype (Fujino et al., 1989), suggesting that *Epi* may represent a constitutive ethylene signalling mutant. In addition, treatment of seedlings with exogenous ethylene enhanced the triple-response phenotype, suggesting that the constitutive ethylene response displayed by *Epi* is not saturated (S. Lee and J. Giovannoni, unpublished).

To date, only one ethylene signal transduction mutant resulting in constitutive ethylene signalling has been identified (*ctr1*; Kieber et al., 1993). *TCTR1* was utilized as an RFLP probe on a subset of 123 mutant F2 progeny derived from a cross between *L. esculentum* cv. VFN8 (*Epi/Epi*) and *L. cheesmannii* (*epi/epi*), in order to test the possibility that the *Epi* mutation may represent a defect in the *TCTR1* gene. The *TCTR1* RFLPs segregated in a 1:2:1 ratio within the subpopulation of mutant F2s, indicating that the *TCTR* locus is not linked to *Epi*. We subsequently have shown that *TCTR1* and *Epi* map to different chromosomes (S. Lee and J. Giovannoni, unpublished). This result suggests that *Epi* represents an ethylene signalling component whose *Arabidopsis* counterpart either has not been identified or does not exist.

Based on the constitutive ethylene response phenotype of *Epi* seedlings and leaves, we predicted that fruit ripening, pedicel abscission, leaf senescence and additional ethylene-mediated developmental processes may be accelerated or exaggerated in the *Epi* mutant. In a preliminary test of ten normal and ten mutant fruit, tagged at anthesis, we did not observe any significant change in time to onset or completion of ripening. One explanation for this result is that developmental regulators required for ripening control this process via modulation of ethylene signal transduction during fruit development. Alternatively, *Epi* may represent a component of ethylene signalling which does not act in the fruit. Consistent with either hypothesis is the observation that all ripening-related and ethylene-inducible genes examined were regulated normally during *Epi* fruit ripening (S. Lee and J. Giovannoni, unpublished). In contrast, the ethylene-inducible chitinase 9 gene (*CH9*; Danhash *et al.*, 1993) was constitutively expressed in both leaves and mature green fruit of the *Epi* mutant, and ethylene inducible in corresponding non-mutant tissues (S. Lee and J. Giovannoni, unpublished). This last result is consistent with the observation of constitutive chitinase expression in the *Arabidopsis ctr1* mutant (Kieber *et al.*, 1993).

In order to differentiate between the two hypotheses above, and to determine epistatic interaction between *Epi* and *Nr*, *Epi/Epi*; *Nr/Nr* double mutants were created. If the first hypothesis is true, one would anticipate that tomato plants harbouring both the *Nr* and *Epi* mutations would result in normally ripening fruit. Conversely, if *Epi* has no role in fruit ethylene signalling, then *Nr* phenotype fruit would be present in the double mutant. Preliminary analysis of *Epi/Epi*; *Nr/Nr* double mutants revealed that seedlings and leaves show a constitutive ethylene response while fruit displayed incomplete ripening characteristic of the *Nr* mutation (H. Yen and J. Giovannoni, unpublished). These results suggest that *Epi* represents a component of ethylene signal transduction that does not operate in all stages of development, and in particular has no role during fruit ripening. We are performing additional experiments to test this hypothesis.

Isolation of Additional Ripening-related cDNAs

A total of 57 ripening-related display PCR products were identified and eluted from acrylamide gels as described in Oh *et al.* (1995). Following re-amplification, each display-cDNA was radiolabelled and hybridized to an RNA gel-blot containing RR, BR, MG and MG + ethylene total RNA (RR, red ripe; BR, breaker; MG, mature green). Twenty-two display-cDNAs demonstrating fruit RNA gel-blot patterns identical to those observed in the corresponding differential display acrylamide gels were recovered and hybridized to whole seedling and leaf total RNA to verify fruit specificity. Most ripening-related display-PCR products showed ethylene inducibility, and several which appeared to be developmentally regulated were also recovered (P. Kannan, B. Oh and J. Giovannoni, unpublished). We refer to these display-cDNAs with the acronym DDTFR (differential display tomato fruit ripening).

GenBank database analysis of 20 of these sequences resulted in the identification of five display-cDNAs which corresponded to previously characterized tomato genes. For example, DDTFR-2 and DDTFR-3 show perfect homology with the 3' end of tomato fruit polygalacturonase, while DDTFR-9 and DDTFR-15 are identical to 3' sequences found within the heat shock-like cDNA, pTOM66 (Gray *et al.*, 1992). Both DDTFR-11 and DDTFR-14 showed greater than 70% sequence identity with two different *Arabidopsis* sequence-tagged site (STS) clones of unknown function. The remaining DDTFR clones showed no significant homologies with any sequences deposited in GenBank, nor with each other. We have isolated full or near full-length cDNAs corresponding to a number of these cDNAs and have identified homologies for several of these ripening-related genes which have not been reported previously. Among these are cDNAs with homology to steroid hormone receptors, transcription factors and translation factors. We have initiated construction of antisense genes to determine the effects of these sequences during fruit development and ripening.

Summary

A number of single and quantitative trait loci influencing development and quality parameters in ripening tomato fruit have been identified and characterized as to their respective effects and, in some cases, corresponding genes have been isolated. Efforts in numerous laboratories, including our own, have focused upon the isolation of specific genes which regulate the ripening process and related fruit quality characters. Together, these efforts have resulted in the isolation of genes involved in numerous aspects of the ripening phenotype including cell wall metabolism, ethylene biosynthesis and perception, pigment biosynthesis and susceptibility to postharvest pathogens. Through analysis of transgenic plants, many of these same genes have been shown to influence more subtle fruit quality characteristics including nutritional composition and processing characteristics.

Specific efforts in our laboratory are focused in three general areas. The first is toward isolation and characterization of genes which represent upstream global developmental regulators of ripening such as the *ripening-inhibitor* (*rin*) and *non-ripening* (*nor*) genes. Progress to date toward map position-based cloning of both loci includes isolation of high molecular weight clones harbouring each, and dissection of the said cloned genomic sequences for ultimate target gene isolation.

Our second focus is on analysis of ethylene signal transduction components and analysis of corresponding gene expression and function during the ripening process. We have isolated a putative tomato homologue of the *Arabidopsis CTR1* gene and have shown that it is ethylene regulated during tomato fruit development. This represents a second component of ripening-related ethylene signal transduction, in addition to the *Never-ripe* (*NR*) ethylene receptor, whose mRNA accumulation is itself under ethylene control. In addition, we have shown that a putative tomato constitutive ethylene response mutant (*Epi*; *Epinastic*) does not represent a mutation in the tomato *CTR1* gene *TCTR1*.

Finally, we are pursuing differential screens to isolate additional ripening-related genes that may represent specific catalytic or regulatory steps contributing to the ripening process. A number of genes previously identified by others were isolated in our screen; however, novel ripening-related genes include those with homology to steroid hormone receptors, transcription factors and translation factors. Molecular and genetic analysis of the said ripening-related genes shows that none are linked to the *rin* or *nor* ripening regulatory loci, and most are ethylene regulated.

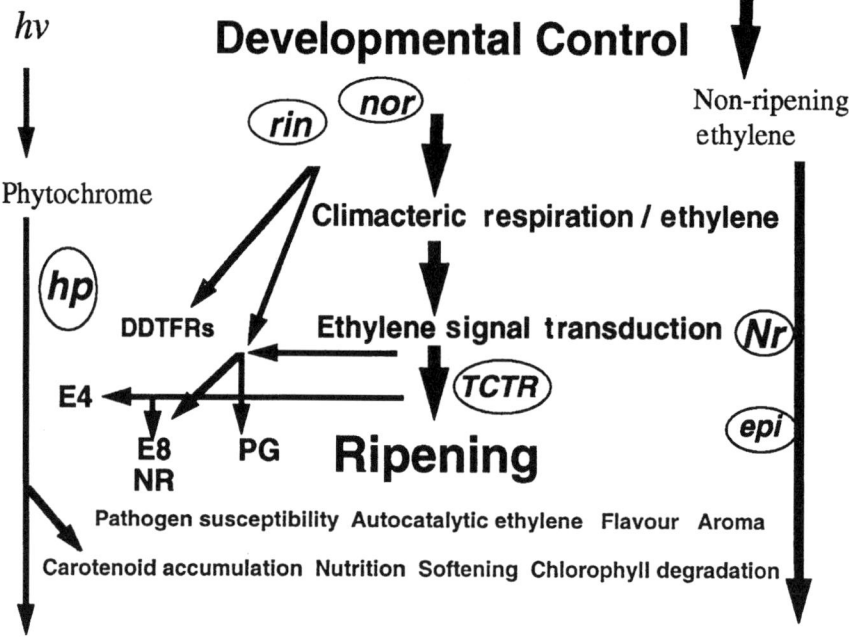

Fig. 1.6. The manifestation of fruit ripening is coordinated by genetic determinants which are themselves influenced by complex interactions among developmental, hormonal and environmental stimuli.

While great strides have been made toward understanding ethylene signal transduction in the model system of *Arabidopsis*, little has been done to characterize corresponding genes in additional species. The need for such effort is exemplified by the fact that the two putative tomato ethylene signalling genes reported to date, *NR* and *TCTR1*, are characterized by unique developmental and ethylene-mediated regulation. This observation suggests that ethylene signal transduction may be modulated, under certain circumstances, by additional layers of developmental, environmental or hormonal regulation to ensure optimal ethylene responses during plant growth and development. As summarized in Fig. 1.6, analysis of tomato ethylene response genes, light signal transduction mutants and developmental regulators suggests numerous layers of regulatory complexity which remain to be explored, and numerous opportunities which remain for genetic manipulation of fruit development, ripening and quality.

References

Abeles, F., Morgan, P. and Saltveit, M. (1992) *Ethylene in Plant Biology*. Academic Press, San Diego.

Alpert, K., Grandillo, S. and Tanksley, S. (1995) *fw 2.2*: a major QTL controlling fruit weight is common to both red- and green-fruited tomato species. *Theoretical and Applied Genetics* 91, 994–1000.

Biggs, M. and Handa, A. (1989) Temporal regulation of polygalacturonase gene expression in fruits of normal, mutant, and heterozygous tomato genotypes. *Plant Physiology* 89, 117-125.

Bleeker, A., Estelle, M., Somerville, C. and Kende, H. (1988) Insensitivity to ethylene conferred by a dominant mutation in *Arabidopsis thaliana*. *Science* 241, 1086-1089.

Danhash, N., Wagemakers, C., van Kan, J. and de Wit, P. (1993) Molecular characterization of four chitinase cDNAs obtained from *Cladisporium fulvum*-infected tomato. *Plant Molecular Biology* 22, 1017-1029.

Deikman, J., Kline, R. and Fischer, R. (1992) Organization of ripening and ethylene regulatory regions in a fruit-specific promoter from tomato (*Lycopersicon esculentum*). *Plant Physiology* 100, 2013-2017.

DellaPenna, D., Alexander, D. and Bennett, A. (1986) Molecular cloning of tomato fruit polygalacturonase: analysis of polygalacturonase mRNA levels during ripening. *Proceedings of the National Academy of Sciences USA* 83, 6420-6424.

DellaPenna, D., Lincoln, J. E., Fischer R. L. and Bennett, A.B. (1989) Transcriptional analysis of polygalacturonase and other ripening associated genes in Rutgers, *rin*, *nor*, and *Nr* tomato fruit. *Plant Physiology* 90, 1372-1377

Ecker, J.R. (1995) The ethylene signal transduction pathway in plants. *Science* 268, 667-675.

Fray, R. and Grierson, D. (1993) Identification and genetic analysis of normal and mutant phytoene synthase genes of tomato by sequencing, complementation, and co-suppression. *Plant Molecular Biology* 22, 589-602.

Fujino, D., Burger, D. and Bradford, K. (1989) Ineffectiveness of ethylene biosynthetic and action inhibitors in phenotypically reverting the *Epinastic* mutant of tomato (*Lycopersicon esculentum* Mill.). *Journal of Plant Growth Regulators* 8, 53-61.

Giovannoni, J. (1993) Molecular and genetic analysis of tomato fruit development and ripening. *Methods in Plant Biochemistry* 10, 251-285.

Giovannoni, J., DellaPenna, D., Bennett, A. and Fischer, R. (1989) Expression of a chimeric polygalacturonase gene in transgenic *rin* (ripening inhibitor) tomato fruit results in polyuronide degradation but not fruit softening. *The Plant Cell* 1, 53-63.

Giovannoni, J., Noensie, E., Ruezinsky, D., Lu, X., Tracy, S., Ganal, M., Martin, G., Pillen, K. and Tanksley, S. (1995) Molecular genetic analysis of the *ripening-inhibitor* and *non-ripening* loci of tomato: a first step in genetic map-based cloning of fruit ripening genes. *Molecular and General Genetics* 248, 195-206.

Gray, J.E., Picton, S., Shabbeer, J., Schuch, W. and Grierson, D. (1992) Molecular biology of fruit ripening and its manipulation with antisense genes. *Plant Molecular Biology* 19, 69-87

Gray, J.E., Picton, S., Giovannoni, J.J. and Grierson, D. (1994) The use of transgenic and naturally occurring mutants to understand and manipulate tomato fruit ripening. *Plant, Cell and Environment* 17, 557-571

Hamilton, A., Lycett, G. and Grierson, D. (1990) Antisense gene that inhibits synthesis of the hormone ethylene in transgenic plants. *Nature* 346, 284-287.

Harriman, R., Tieman, D. and Handa, A. (1991) Molecular cloning of tomato pectin methylesterase gene and its expression in Rutgers, ripening inhibitor, nonripening and *Never ripe* tomato fruits. *Plant Physiology* 97, 80-87.

Hobson, G. and Grierson, D. (1993) Tomato. In: Seymour, G.B., Taylor, J.E. and Tucker, G.A. (eds), *Biochemistry of Fruit Ripening*. Chapman and Hall, London, pp. 405-442.

Kieber, J., Rothenberg, M., Roman, G., Feldman, K. and Ecker, J. (1993) *CTR1*, a negative regulator of the ethylene response pathway in *Arabidopsis*, encodes a member of the Raf family of protein kinases. *Cell* 72, 427-441.

Koshland, D. (1995) The two-component pathway comes to eukaryotes. *Science* 262, 532.

Kramer, M., Sanders, R., Sheehy, R., Melis, M., Kuehn, M. and Hiatt, W. (1990) Field evaluation of tomatoes with reduced polygalacturonase by antisense RNA. In: Bennett, A. and O'Neill, S. (eds), *Horticultural Biotechnology*. Alan R. Liss, New York, pp. 347-355.

Lanahan, M.B., Yen, H.C., Giovannoni, J.J. and Klee, H.J. (1994) The *Never Ripe* mutation blocks ethylene perception in tomato. *The Plant Cell* 6, 521-530.

Lehman, A., Black, R. and Ecker, J. (1996) *HOOKLESS1*, an ethylene response gene, is required for differential cell elongation and de-etiolation in the *Arabidopsis* hypocotyl. *Cell* 85, 183-194.

Lincoln, J., Cordes, S., Read, E. and Fischer, R. (1987) Regulation of gene expression by ethylene during *Lycopersicon esculentum* (tomato) fruit development. *Proceedings of the National Academy of Sciences USA* 84, 2793-2797.

Mattoo, A.K. and Suttle, J.C. (1991) *The Plant Hormone Ethylene*. CRC Press, Boca Raton, Florida.

Montgomery, J., Pollard, V., Deikman, J. and Fischer, R. (1993) Positive and negative regulatory regions control the spatial distribution of polygalacturonase transcription in tomato fruit pericarp. *The Plant Cell* 5, 1049-1062.

Oeller, P.W., Wong, L.M., Taylor, L.P., Pike, D.A. and Theologis, A. (1991) Reversible inhibition of tomato fruit senescence by antisense 1-aminocyclopropane-1-carboxylate synthase. *Science* 254, 427-439.

Oh, B. J., Balint, D. and Giovannoni, J. (1995) A modified procedure for PCR-based differential display and demonstration of use in plants for isolation of fruit ripening-related genes. *Plant Molecular Biology Reporter* 13, 70-81.

Paterson, A., Lander, E., Hewitt, J., Peterson, S., Lincoln, S. and Tanksley, S. (1988) Resolution of quantitative traits into Mendelian factors using a complete linkage map of restriction fragment length polymorphisms. *Nature* 335, 721-726.

Paterson, A., Damon, S., Horowitz, J., Zamir, D., Rabinowitch, H., Lincoln, S., Lander, E. and Tanksley, S. (1991) Mendelian factors underlying quantitative traits in tomato: comparison across species, generations, and environments. *Genetics* 127, 181-197.

Picton, S., Gray, J.E., Barton, S.L., AbuBaker, U., Lowe, A. and Grierson, D. (1993) cDNA cloning and characterization of novel ripening-related mRNAs with altered patterns of accumulation in the ripening inhibitor (rin) tomato ripening mutant. *Plant Molecular Biology* 23, 193-207.

Schaller, G. and Bleeker, A. (1995) Ethylene-binding sites generated in yeast expressing the *Arabidopsis ETR1* gene. *Science* 270, 1809-1811.

Sheehy, R., Pearson, J., Brady, C. and Hiatt, W. (1987) Molecular characterization of tomato fruit polygalacturonase. *Molecular and General Genetics* 208, 30-36.

Slater, A., Maunders, M., Edwards, K., Schuch, W. and Grierson, D. (1985) Isolation and characterization of cDNA clones for tomato polygalacturonase and other ripening-related proteins. *Plant Molecular Biology* 5, 137-147.

Smith, C., Watson, C., Ray, J., Bird, C., Morris, P., Schuch, W. and Grierson, D. (1988) Antisense RNA inhibition of polygalacturonase gene expression in transgenic tomatoes. *Nature* 334, 724-726.

Tanksley, S. and Nelson, J. (1996) Advanced backcross QTL analysis: a method for the simultaneous discovery and transfer of valuable QTLs from unadapted germplasm into elite breeding lines. *Theoretical and Applied Genetics* 92, 191-203.

Tanksley, S., Ganal, M., Prince, J., deVincente, C., Bonierbale, M., Broun, P., Fulton, T., Giovannoni, J., Grandillo, S., Martin, G., Messeguer, R., Miller, J., Miller, L., Paterson,

A., Pineda, O., Roder, M., Wing, R., Wu, W. and Young, N. (1992) High density molecular maps of the tomato and potato genomes. *Genetics* 132, 1141-1160.

Tanksley, S., Grandillo, S., Fulton, T., Zamir, D., Eshed, Y., Petiard, V., Lopez, J. and Beck-Bunn, T. (1996) Advanced backcross QTL analysis in a cross between an elite processing lone of tomato and its wild relative *L. pimpinellifolium. Theoretical and Applied Genetics* 92, 213-224.

Theologis, A. (1992) One rotten apple spoils the whole bushel: the role of ethylene in fruit ripening. *Cell* 70, 181-184.

Theologis, A., Oeller, P., Wong, L., Rothmann, W. and Gantz, D. (1993) Use of a tomato mutant constructed with reverse genetics to study fruit ripening, a complex developmental process. *Developmental Genetics* 14, 282-259.

Tigchelaar, E.C., McGlasson, W.B. and Buescher, R.W. (1978) Genetic regulation of tomato fruit ripening. *HortScience* 13, 508-513.

Ursin, V. (1987) Morphogenetic and physiological analyses of two developmental mutants of tomato, *Epinastic* and *diageotropica*. PhD Dissertation, University of California, Davis, California.

Wilkinson, J., Lanahan, M., Yen, H., Giovannoni, J. and Klee, H. (1995) An ethylene-inducible component of signal transduction encoded by *Never-ripe. Science* 270, 1807-1809.

Yelle, S., Chetelat, R., Dorais, M., DeVerna, J. and Bennett, A. (1991) Sink metabolism in tomato fruit. IV. Genetic and biochemical analysis of sucrose accumulation. *Plant Physiology* 95, 1026-1035.

Yen, H., Lee, S., Tanksley, S., Lanahan, M., Klee, H. and Giovannoni, J. (1995) The tomato *Never-ripe* locus regulates ethylene-inducible gene expression and is linked to a homologue of the *Arabidopsis ETR1* gene. *Plant Physiology* 107, 1343-1353.

Yen, H., Shelton, A., Howard, L. Vrebalov, J. and Giovannoni, J. (1997) The tomato *high pigment (hp)* locus maps to chromosome 2 and influences plastome copy number and fruit quality. *Theoretical and Applied Genetics* 95, 1069-1079.

2 Improving Tomato Fruit Quality by Cultivation

L.C. Ho

Horticulture Research International, Wellesbourne, Warwick CV35 9EF, UK

Introduction

There has been a steady increase in the annual yield of glasshouse tomatoes in the UK over the last 26 years (Fig. 2.1). The average yield from heated glasshouses in 1996 was 373 t ha^{-1}, which was achieved by an annual increase of more than 6% over the period. Some of this increase arose from the introduction of cultivars resistant to pests and diseases, but most was achieved by extending the cropping period and by increasing the fruit load per plant and per cropping area. In addition, the proportion of marketable yield has benefited greatly from improved plant nutrition, resulting from hydroponic culture and from the application of integrated biological–chemical management on diseases and pests. The successful application of research findings in plant physiology and nutrition has also played a major part in these improvements in yield and quality, for example through improved temperature regimes, CO_2 enrichment and highwire canopy stands. The difference in yield achieved by the best growers (up to 600 t ha^{-1}) compared with the national average can be attributed largely to the level of implementation of these cultivation practices.

While improvements in yield over the last 26 years have enabled the British tomato industry to maintain a present share of 28% in the domestic market, the industry now has to improve fruit quality yet further to resist a new wave of competition from the southern European growers. As consumer choice becomes ever more sophisticated, the future of the British tomato industry will depend on whether it will provide a reliable supply of highest quality product, while remaining responsive to new opportunities arising from changing consumer preferences. The emphasis for future research in plant physiology and nutrition should, therefore, shift even further towards establishing the scientific basis of fruit quality, rather than focusing on yields.

© CAB INTERNATIONAL 1998. *Genetic and Environmental Manipulation of Horticultural Crops* (eds K.E. Cockshull, D. Gray, G.B. Seymour and B. Thomas)

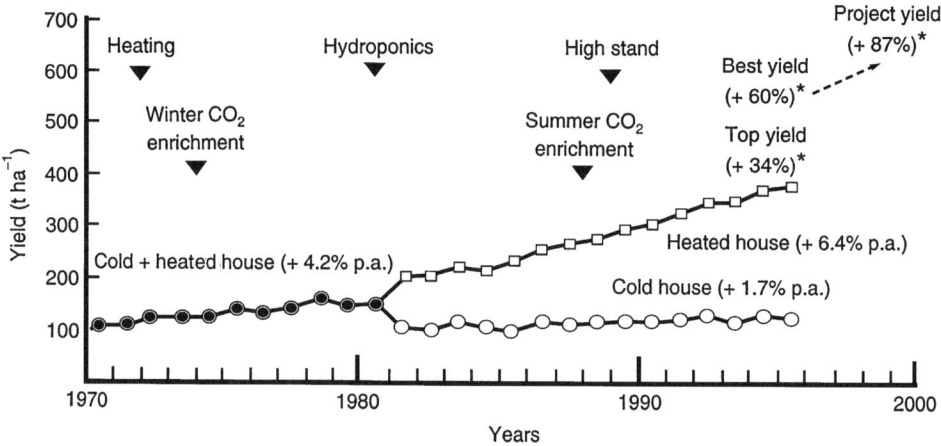

Fig. 2.1. The average annual yield of glasshouse tomato in the UK between 1970 and 1996 (based on UK MAFF statistics) and the introduction of cultivation practice over the same period.

Tomato Fruit Quality Criteria

The consumer's preference for certain types of tomato is likely to be dictated by perceptions of appearance, taste/texture and nutritional value, as well as by the cost of the product (Table 2.1). The appearance of the fruit (size, shape, colour) has the most immediate and profound effect on consumer choice and, accordingly, produce for fresh market is graded principally on the appearance of the fruit. For this reason, growers are very much preoccupied with the prevention of visible defects. Defects depend partly on cultivar but they also vary with nutritional and environmental conditions during cultivation and storage. More persistent factors determining the consumer's choice are taste and texture of the fruit, particularly for fresh salads. The familiar sweet-and-sour taste of tomato is determined principally by the concentrations of sugars (i.e. glucose, fructose and sucrose) and of acids (i.e. malate and citric acids) in the fruit juice (see Davies and Hobson, 1981). The composition of sugars and acids within the fruit depends on cultivar and on light and temperature conditions during growth (which sets the rates of biosynthesis and metabolism within the fruit; see Hobson and Davies, 1971; Ho, 1996), while the concentration of sugars and acids in the juice can also be affected directly by fruit water relationships (Adams and Ho, 1989; Mitchell *et al.*, 1991). However, to encourage the industry to develop and market fruit of improved flavour, such fruit would have to command a significant price premium. Tomato also offers added value because it is rich in vitamins A and C (Davies and Hobson, 1981), as well as in dietary antioxidants such as carotenes (especially lycopene). The latter may act to reduce damage to body tissues from chemical by-products known as free radicals, and have been linked with the treatment of various ailments including certain cancers (see Gerster, 1997). The levels of vitamins and lycopene in the

Table 2.1. The perception of consumers, retailers and growers of different aspects of tomato fruit quality.

Quality	Grower's challenge	Retailer's profit	Consumer's preference
Appearance/texture			
Colour	Sun scale/uneven ripening		Uniform/bright
Size	Grading		Uncritical
Shape	Boxy		Round
Firmness	Handling	Shelf life	Flavour releases
Skin	Netting/cracking	Shelf life	Shininess
Taste/aroma	Sugar/acid	Premium	Sweet/sour
	Volatile		Fragrance
Health/nutrition	Lycopene	Premium	Diet
	Vitamin A/C		

fruit can be affected by temperature and by the nutritional status of the fruit (Grierson and Kader, 1986). Therefore, consumer preference for tomato potentially could be improved by the adoption of cultivation and storage conditions which optimize the nutritional value of the fruit. Furthermore, various fruit quality problems, such as blossom-end rot (BER), blotchy ripening and softening, are exacerbated by adverse environmental conditions during growth and storage. Clearly then, cultivation and storage conditions are vital to almost all aspects of tomato fruit quality.

The Physiological Basis for Improving Fruit Quality

In order to improve fruit quality with existing cultivars, we must first identify those physiological processes responsible for the quality attributes, and those disorders responsible for the defects. The following examples illustrate how different aspects of fruit quality can be improved by cultivation.

Dry matter content and sweetness

High dry matter content (i.e. the percentage of fruit fresh weight as dry matter) or high soluble solute content (i.e. the percentage of solute weight in the fruit juice) of a tomato fruit is highly desirable for both the processing and the fresh market industry. For processing tomato, high dry matter content means lower cost in concentrating the fruit content into paste and higher quality in the paste and juice. Sugars account for about half of the fruit dry matter and over 65% of the soluble solute in the fruit juice (Hobson and Davies, 1981), and the sweetness of tomato is determined by the sugar concentration. Modern cultivars tend to have a dry matter content of between 5 and 7.5%, but this can also be manipulated readily by cultivation.

The dry matter in a tomato fruit consists largely of photoassimilates imported during fruit development; only a small fraction is produced by carbon fixation within the fruit (see Ho and Hewitt; 1986; Ho, 1996). As the photoassimilate and most of the water were imported by the phloem conducting tissue, both the absolute amount and the proportions of dry matter and water in the fruit are reflections of the volume and concentration of the phloem sap (Ho et al., 1987). Light interception by the crop canopy ultimately determines the fruit yield (Cockshull et al., 1992), but the import of assimilate by individual fruit is determined by the partitioning of photoassimilate in the plant (Ho, 1984). Therefore, the accumulation of dry matter by an individual fruit is related to the change of canopy photosynthesis through the seasons (Winsor and Adams, 1982) and can also be affected by the leaf/fruit ratio in the plant (Davies and Winsor, 1967). In practice, supplementary lighting in the winter can increase fruit dry matter content, while partial deleafing decreases fruit dry matter content. By contrast, winter CO_2 enrichment, although it increased yield, did not significantly affect the dry matter content of the fruit in previous works (see van de Vooren et al., 1986).

Dry matter content of a tomato fruit is relatively low, as about 93-95% of the fresh weight is water (Ho et al., 1987). Under a given light regime, the dry matter content of tomato fruit can be manipulated effectively by irrigation. By raising the salinity in the feed either by increasing the concentration of the nutrients or by adding NaCl, less water is accumulated by the fruit and the dry matter content is thus increased throughout fruit development (Ehret and Ho, 1986a). Fruit dry matter content has been shown to be related linearly to the electrical conductivity (i.e. EC) of the feed: dry matter content increases by about 0.23% per $mS\ cm^{-1}$, over the range of $2-10\ mS\ cm^{-1}$ in NFT culture (Fig. 2.2). Fruit dry matter content can thus be manipulated predictably and accurately. Thus both fruit sweetness and dry matter content are amenable to manipulation through the light conditions and water supply. It is anticipated that further improvement in dry matter content will be achieved by using supplementary lighting, fruit pruning, CO_2 enrichment and EC treatments in proper combination.

Uniform fruit size

Variation of average fruit size throughout the season probably originates from differences in fruit number compared with assimilate supply to the individual fruit. In the winter/springtime when canopy photosynthesis is limited by low light, the supply of assimilate to all the fruit set is not sufficient to sustain the growth of the majority of fruits to the preferred size (i.e. Class I, size D: 47-57 mm in diameter or 50-90 g) for round tomato. Consequently, more than half of the fruit are in size E (i.e. 40-47 mm) in the early crop. On the other hand, the proportion of fruit larger than size D (i.e. size C; larger than 57 mm) increases in

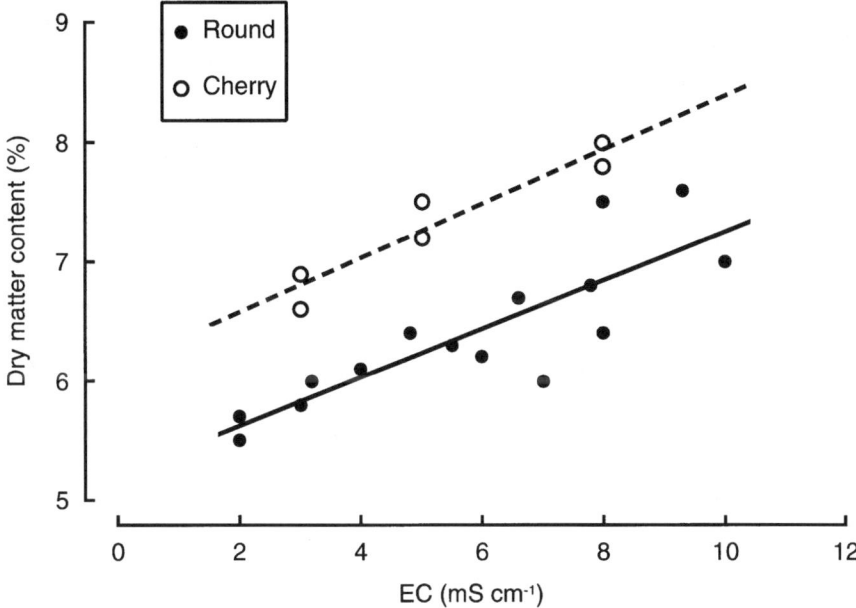

Fig. 2.2. Relationships between EC in the feed and the dry matter content in tomato fruit grown in nutrient-film technique. Data for the round tomato are from Massey *et al.* (1984), Ehret and Ho (1986a) and Adams and Ho (1989); data for the cherry tomato are from Gough and Hobson (1990).

the summer, when light is no longer limiting. For retailers, uniform fruit size throughout the seasons is highly desirable.

Both canopy photosynthesis and fruit number within the canopy are affected by plant density. For long-season crops, a planting density of 10,000 per acre (i.e. 2.5 plants m^{-2}) was used commonly to achieve high yield. However, in the early season the majority of fruit were small and only 35% fruit yield was size D. We demonstrated that the proportion of tomato fruit in the preferred size class could be increased by altering the shoot density (i.e. sideshoot taking) or the fruit number (i.e. truss thinning) (Cockshull and Ho, 1995); for example, starting with a lower planting density of 8000 per acre and then taking sideshoots in alternate plants, so that shoot density increased to 12,000 per acre later in the season. Over 70% of the early fruit yield was size D (with far less size E), while more than 80% of later yield was also size D (with less in size C). The proportion of early yield in size D can be increased further by pruning to reduce the fruit number from nine to six per truss. Thus, by better matching of fruit production to assimilate supply throughout the season, a significant proportion of fruit can be promoted into the preferred size class.

Application of these findings over the last few years has enabled the British tomato industry to produce a much more uniform and cost-effective product than formerly. It is anticipated that further improvement can be made using the same approach. For example, the frequent late-season increase in size class E could be remedied by reducing canopy size (to reduce shading) and by truss pruning (to reduce fruit number) during the period when light levels are declining.

Summer fruit defects

There are a number of fruit defects (e.g. dark patch, netting, softening, blotchy ripening and goldspot) which commonly occur in the summer when light levels and temperature are high and humidity is low. Preliminary results demonstrated that these defects may be caused by high temperature interacting with nutrient supply (Fussell and Ho, 1997). Different temperature regimes were applied in the glasshouse, together with different levels of Ca and K in the feed. It was found that the incidence of defects such as dark patch, softness and blotchy ripening were significantly enhanced by raising the glasshouse temperature (venting set point increased from 21 to 25°C). However, these defects were not affected by the level of K (i.e. 400 vs. 230 p.p.m.) or Ca (i.e. 250 vs. 100 p.p.m.) in the feed.

Softening of a tomato fruit is a consequence of the normal ripening processes. However, the ripe fruit in the summer tend to be too soft for normal handling, causing increased wastage for the retailers. In recent years, long-life cultivars such as 'Daniella' with delayed ripening have been introduced. These initially were welcomed by the processing industry and the retailers for the fresh market, but acceptance by the consumers has been declining. Optimizing the growing conditions to improve the firmness of existing good flavour cultivars is required. Therefore, the possible roles of cell turgor, cell wall structure and tissue strength in fruit firmness have to be defined, and the effects of temperature, water relationships and fruit nutrient status on these parameters have to be quantified.

Blotchy ripening, by contrast, is a ripening disorder. The incidence of this defect is enhanced by high temperature (25 vs. 21°C venting set point) and low potassium (230 vs. 400 p.p.m. in the feed), but its cause is still very poorly understood. We should ascertain whether the failure of biosynthesis of lycopene in certain areas is the principal cause of this defect, and how the process is affected by fruit temperature and potassium status in the fruit tissue.

BER and goldspot

Both blossom-end rot (BER) and goldspot in tomato are physiological disorders related to calcium status. BER is caused by Ca deficiency in the distal fruit tissue during rapid fruit expansion (see Sheer, 1975; Bangerth, 1979), while goldspot is caused by excess deposition of calcium oxalate crystals just under the fruit skin during ripening (Den Outer and van Veenendaal, 1988; De Kreij *et al.*, 1992). BER causes much greater economic loss than goldspot because BER reduces the marketable yield. There is a wide range of susceptibility to BER among modern cultivars (Adams and Ho, 1992; Ho *et al.*, 1995), and the cultivars susceptible to BER tend to be resistant to goldspot and *vice versa* (Nukaya *et al.*, 1995). Apart from cultivar difference, the incidence of BER can be enhanced by certain growing conditions. As the BER-susceptible cultivars tend to be good yielders, optimizing growing conditions to reduce BER has been found to be effective with these cultivars.

Prevention of BER

We have investigated the effects of certain growing conditions on the Ca status in the rapidly expanding fruit and identified four common causes of BER: (i) low Ca status in the plant; (ii) low Ca transport to the distal fruit tissue; (iii) extra demand for Ca for accelerated fruit expansion; and (iv) effect of an imbalance between Ca and other nutrients (e.g. N and P) on the cell membrane permeability. Accordingly, a range of cultural measures have been formulated to prevent the induction of BER.

Optimizing the absorption of Ca by roots

The ultimate cause of BER is Ca deficiency in the fruit which can be induced by temporary withdrawal of Ca supply (Adams and ElGizawy, 1988) or by insufficient supply of Ca (i.e. < 100 p.p.m. in the feed; Massey et al., 1984). A supply of 200 p.p.m. Ca in the feed is the recommended practice in current hydroponic cultures. However, the uptake of Ca by roots is essentially proportional to water uptake (Ho et al., 1995), and water uptake is affected, through canopy transpiration, by light and humidity (Aikman and Houter, 1990) and by root temperature (Adams, 1989). Thus, the plant Ca status can be altered through growing conditions, as well as by changing the Ca level in the feed. Apart from Ca stress, the plant Ca status frequently is reduced by salinity in the root zone (i.e. osmotic stress; Ehret and Ho, 1986c). In practice, using a high concentration of nutrients or saline water in the feed resulting in high EC would greatly reduce the uptake of Ca despite a sufficient supply of Ca in the feed. The importance of root function in Ca uptake has also been demonstrated in that the Ca uptake can be reduced by poor aeration in the root zone, resulting in a higher incidence of BER (Tachibana, 1988). Therefore, even where there is sufficient Ca in the feed, BER incidence can be reduced if high EC (> 5 mS cm^{-1}), high NH_4 (> 10% total nitrogen supply; Massey and Winsor, 1980), poor aeration (prolonged water logging in the rockwool system) and extreme temperature (< 14°C or > 30°C) in the root zone can be avoided.

Improving the distribution of Ca to rapidly expanding fruit

The transport of Ca from roots to fruit in tomato is intrinsically low because Ca moves with the transpiration stream, and the fruit has only very low rates of transpiration (Ehret and Ho, 1986b). Consequently, the Ca concentration in the fruit is lower than in any other organs in the shoot (Ho and Adams, 1994). The Ca concentration of the fruit may be reduced even further if the canopy transpiration is high because more of the Ca is then diverted away from fruit to the transpiring leaves (Adams and Holder, 1992). Furthermore, the distribution of Ca in the fruit is not uniform. The xylem network, which carries the transpiration stream, is very poorly developed towards the distal end of the fruit, thus very little of the transpiration stream reaches this region (Ehret and Ho,

1986b). Consequently, the Ca concentration in the distal pulp tissue is lowest in the fruit (Adams and Ho, 1992). The xylem network in the distal part of the fruit is especially poorly developed in the susceptible cultivars, and its development is reduced still further if the feed solution has high salinity (Belda and Ho, 1993; Belda *et al.*, 1996). Therefore, high salinity not only reduces the uptake of Ca by the roots, but it also reduces the transport of Ca into the distal tissue. Furthermore, the Ca concentration of the fruit decreased substantially during the period of rapid fruit expansion (about 2 weeks after anthesis; Ehret and Ho, 1986c). Consequently, the Ca concentration of the distal fruit tissue may become critically low during the period of rapid expansion. This is the critical time when BER most commonly is induced. When the Ca concentration in the distal pulp tissue is reduced below the critical level for cell membrane permeability, BER will develop.

To improve the distribution of Ca to the rapidly expanding fruit, canopy transpiration has to be reduced in order to divert the Ca flow from the leaf canopy. By reducing the vapour pressure deficit (VPD) in the glasshouse from 0.8 to 0.1 kPa, the Ca concentration in the fruit has been increased, but the incidence of BER in the cultivars investigated (which were not susceptible to BER) has not been found to be affected by humidity treatments in this range (Adams and Holder, 1992). However, whether the canopy transpiration would be affected more markedly by air flow requires further investigation. As a preventive measure, high VPD (>0.5 kPa) and high EC (>5 mS cm^{-1}) should be avoided.

Regulation of fruit growth rate to provide a better balance between assimilate import and Ca deposition for normal cell enlargement

Independently of the cultivar susceptibility to BER, the seasonal pattern of BER is closely related to changes in irradiation and temperature throughout the season (Ho *et al.*, 1993). The promotion of BER by increasing light and temperature may not be caused solely by effects on the uptake or distribution of Ca in the plant alone, but may also involve effects on the rate of fruit expansion. Fruit growth rate will be faster when more assimilate is available (at higher light levels), and when fruit metabolic activity is higher (at higher temperature) (Pearce *et al.*, 1993). Ca is required for the integrity of both membrane and cell wall, thus the demand for Ca within the fruit will be greatest during accelerated fruit expansion. However, the import of Ca by fruit is not necessarily enhanced by conditions which promote fruit expansion and, consequently, accelerated fruit growth may lead to critically low Ca levels and ensuing BER.

The possible role of accelerated fruit growth in the induction of BER has also been demonstrated by the frequent occurrence of BER when a sunny spell follows a long period of dull weather (Ho and Adams, 1993). Furthermore, the incidence of BER can also be increased by truss pruning (i.e. lowering the fruit/ leaf ratio), as the fewer fruit have a higher rate of fruit expansion (DeKock *et al.*, 1982). Although the critical rate of fruit expansion for the induction of BER has not been quantified, growing conditions for high yield (i.e. high light, temperature and CO_2) should be optimized, so as to avoid bursts of unusually high fruit growth rate, which would be likely to induce BER.

Improving Tomato Quality by Cultivation

Maintaining the balance between Ca and other nutrients for cell enlargement

Although BER is caused principally by local Ca deficiency, there is evidence that BER can be enhanced by nutritional imbalance in the fruit. For instance, the incidence of BER can be enhanced by lowering the P supply from 30 to 5 p.p.m., even where the Ca supply is not altered. To a lesser extent, the incidence of BER can also be reduced by reducing the N-nitrate from 240 to 120 p.p.m. in the feed (Ho and Hand, 1996). As the total Ca uptake and fruit Ca concentration is not affected by either P or N levels in the feed, the leaking of the cells which underlies the symptoms of BER may be caused by low P in the cell membrane or by some interaction of Ca and P or N on the permeability of the membrane. For a better understanding of the possible roles of these nutrients in the induction of BER, improved measurements of Ca, N and P distribution within the fruit are required. The measurements currently available are mostly of levels of these elements in the bulked tissue, and do not permit detailed dissection of their physiological function in the induction of BER. Until the critical concentrations of relevant Ca compounds, such as Ca phosphate (Minamide and Ho, 1993), lipid and phosphate in the cell membrane required for the cell permeability are quantified, the cause of BER will not be fully understood.

Our systematic investigations on the mechanisms of the root uptake of Ca, the distribution of Ca in the plant or inside the fruit, as well as the deposition of Ca in different fruit tissues during the rapid growth period have revealed multiple causes of BER and have provided the physiological basis for BER prevention (Fig. 2.3). By manipulating the growing conditions to avoid nutrient stress, water stress and osmotic stress in the root zone, the uptake of Ca and thus

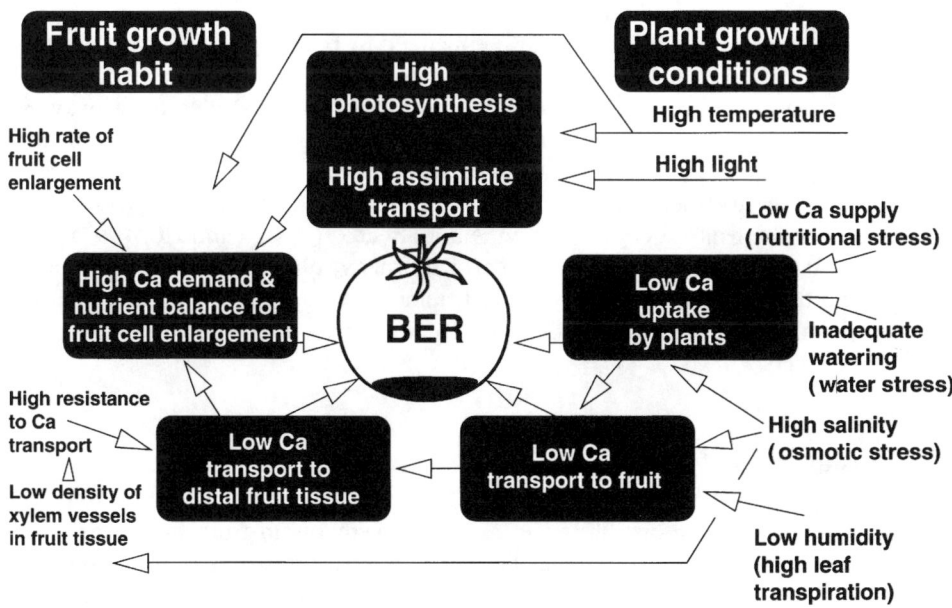

Fig. 2.3. The cause and prevention of BER in tomato.

the Ca status of the plant can be improved. The transport of Ca to the distal fruit tissue can also be improved if osmotic stress in the root zone and excessive canopy transpiration are avoided. Consequently, the Ca concentration of the susceptible tissue (i.e. distal pulp) in the rapidly expanding fruit should not fall below the critical level.

Prevention of goldspot

Goldspot is one of the mild fruit defects in the summer caused by accumulation of excessive Ca oxalate crystals. The principle of prevention is to reduce the deposition of Ca oxalate during fruit ripening.

Optimizing the Ca uptake by roots

The primary cause of goldspot is the abundant supply of Ca in the feed. This is used commonly in hydroponic culture to ensure absence of Ca deficiency. We have found that by reducing the supply of Ca from 250 to 120 p.p.m., the incidence of goldspot was greatly reduced (Fussell and Ho, 1997). The incidence of goldspot can also be reduced by increasing the supply of NO_3, NH_4 (Nukaya *et al.*, 1995) or K (Fussell and Ho, 1997), or by raising the salinity (Sonneveld and Voogt, 1990), but all these practices have to be applied with care as they reduce the root uptake of Ca and potentially can lead to enhanced BER. Therefore, an optimum Ca supply in the feed with a root environment favourable for root function, such as good aeration and moderate temperature, will ensure sufficient Ca uptake for fruit growth but without incurring problems of excess.

Reducing the deposition of Ca as Ca oxalate

The incidence of goldspot is also enhanced by both high humidity (De Kreij *et al.*, 1992) and high temperature (Fussell and Ho, 1997). Since high humidity decreases canopy transpiration, and canopy transpiration diverts Ca away from the fruit (Adams and Holder, 1992), the higher incidence of goldspot under these conditions could be a direct consequence of higher fruit Ca content. However, the incidence of goldspot can be increased by high temperature even when the fruit Ca concentration is not affected (Fussell and Ho, 1997). There must, therefore, be some additional or alternative direct effect of temperature on the deposition of calcium oxalate leading to goldspot. By avoiding high temperature and excessive supply of Ca, the incidence of goldspot should be reduced.

Conclusions

The examples above illustrate how improvements in fruit quality, as well as reduction in fruit defects, may be achieved by manipulating both the shoot and root environment. In the quest for better fruit quality, research is required both on providing better cultivars and on optimizing cultivation conditions to realize the potential of the quality. While the growing environment can be manipulated to enhance some of the attributes of quality or to eliminate some defects, new cultivars having higher sugars and acids in fruit and better texture or better xylem transport capacity inside the fruit are urgently needed.

The advance of computer control technology for glasshouse crop production enables growers readily to manipulate the growing environments for better fruit quality. However, the benefits of such control can be realized fully only when the software for computer control is based on a sound understanding of the physiological responses to the various environmental changes. However, quality criteria change with market demand. Also, the causes of defects are interlinked and interacting. Therefore, applied research and basic research should be well coordinated to ensure a versatile technology for the industry.

Successful technology transfer has been a key contributor to the improvement of UK glasshouse tomato yield in the recent past. The continuing success of technology transfer in this industry should be assured because the industry is now closely involved in funding research. Growers are now involved from the early stages of research projects, and often they have first-hand understanding of the principles underlying new developments. This should lead to even more efficient technology transfer in the future.

Acknowledgements

I dedicate this paper to all my colleagues, who inspired and supported me during my 27 years service in the Glasshouse Crops Research Institute (GCRI)/Horticulture Research International (HRI), for their considerable contribution to the improvement of the yield and quality of glasshouse tomato crop in the UK. All the work reported here was funded by the Biotechnology and Biological Science Research Council (BBSRC), the Ministry of Agriculture, Fisheries and Food (MAFF) and the Horticultural Development Council (HDC) in the UK. Comments from David Gray and Mike Malone on this manuscript were greatly appreciated.

References

Adams, P (1989) Some effects of root temperature on the growth and nutrient uptake of tomatoes in NFT. In: *Proceedings of the 7th International Congress of Soilless Culture, Flevohof, 1988,* pp. 73-82.

Adams, P. and ElGizawy, A.M. (1988) Effect of calcium stress on the calcium status of tomatoes grown in NFT. *Acta Horticulturae* 222, 15-22.

Adams, P. and Ho, L.C. (1989) Effects of constant and fluctuating salinity on the yield, quality and calcium status of tomatoes. *Journal of Horticultural Science* 64, 725-732.

Adams, P and Ho, L.C. (1992) The susceptibility of modern tomato cultivars to blossom-end rot in relation to salinity. *Journal of Horticultural Science* 67, 827-839.

Adams, P. and Holder, R. (1992) Effects of humidity, Ca and salinity on the accumulation of dry matter and Ca by the leaves and fruit of tomato (*Lycopersicon esculentum*). *Journal of Horticultural Science* 67, 137-142.

Aikman, D.P. and Houter, G. (1990) Influence of radiation and humidity on transpiration:

implications for calcium levels in tomato leaves. *Journal of Horticultural Science* 65, 245-253.

Bangerth, F. (1979) Calcium-related physiological disorders of plant. *Annual Review of Phytopathology* 17, 97-122.

Belda, R.M. and Ho, L.C. (1993) Salinity effects on the network of vascular bundles during tomato fruit development. *Journal of Horticultural Science* 68, 557-564.

Belda, R.M, Fenlon, J. and Ho, L.C. (1996) Salinity effects on the xylem vessels in the tomato fruit among cultivars with different susceptibilities to blossom-end rot. *Journal of Horticultural Science* 71, 173-179.

Cockshull, K.E. and Ho, L.C. (1995) Regulation of tomato fruit size by plant density and truss thinning. *Journal of Horticultural Science* 70, 395-407.

Cockshull, K.E., Graves, C.J. and Cave, R.J. (1992) The influence of shading on yield of glasshouse tomatoes. *Journal of Horticultural Science* 67, 11-24.

Davies, J.N. and Hobson, G.E. (1981) The constituents of tomato fruit: the influence of environment, nutrition and genotype. *CRC Critical Reviews of Food Science and Nutrition* 15, 205-280.

Davies, J.N. and Winsor, G.W. (1967) The composition of tomato fruit. *GCRI Annual Report for 1966,* pp. 65-68.

DeKock, P.C., Inkson, R.H.E. and Hall, A. (1982) Blossom-end rot of tomato as influenced by truss size. *Journal of Plant Nutrition* 5, 57-62.

De Kreij, C., Janse, J., van Goor, J. and van Doesburg, J.D.J. (1992). The incidence of calcium oxalate crystals in fruit walls of tomato (*Lycopersicon esculentum* Mill.) as affected by humidity, phosphate and calcium supply. *Journal of Horticultural Science* 67, 45-50.

Den Outer, R.W. and van Veenendaal, W.L.H. (1988). Gold speckles in tomato fruits (*Lycopersicon esculentum* Mill.). *Journal of Horticultural Science* 63, 645-649.

Ehret, D.L. and Ho, L.C. (1986a) Effects of salinity on dry matter partitioning and fruit growth in tomatoes grown in nutrient film culture. *Journal of Horticultural Science* 61, 361-367.

Ehret, D.L. and Ho, L.C. (1986b) Effect of osmotic potential in nutrient solution on diurnal growth of tomato fruit. *Journal of Experimental Botany* 37, 1297-1302.

Ehret, D.L. and Ho, L.C. (1986c) Translocation of calcium in relation to tomato fruit growth. *Annals of Botany* 58, 679-688.

Fussell, M. and Ho, L.C. (1997) Summer fruit quality in tomatoes. *HDC Project News*, 43, 12-13.

Gerster. H. (1997) The potential role of lycopene for human health. *Journal of the American College of Nutrition.*16, 109-126.

Gough, C. and Hobson, G.E. (1990) A comparison of the productivity, quality, shelf-life characteristics and consumer reaction to the crop from cherry tomato plants grown at different levels of salinity. *Journal of Horticultural Science* 65, 431-439.

Grierson, D. and Kader, A.A. (1986) Fruit ripening and quality. In: Atherton, J.G. and Rudich, J. (eds), *The Tomato Crop.* Chapman and Hall, London, pp. 240-280.

Ho, L.C. (1984) Partitioning of assimilates in fruiting tomato plants. *Plant Growth Regulation* 2, 277-285.

Ho, L.C. (1996) Tomato. In: Zamski, E. and Schaffer, A. (eds), *Photoassimilate Distribution in Plant and Crops: Source-Sink Relationship.* Marcel Dekker, New York, pp. 709-728.

Ho, L.C. and Adams, P. (1993) Can we banish the rot? *Grower* 119, 13-18.

Ho, L.C. and Adams, P. (1994) The physiological basis for high fruit yield and susceptibility to calcium deficiency in tomato and cucumber. *Journal of Horticultural Science* 69, 367-376.

Ho, L.C. and Hand, D.J. (1996) Cultivar and cultural aspects of the prevention of blossom-end rot in tomato and pepper. In: *Proceedings of 9th International Congress of Soilless Culture.* Grafoprint, Bennekom, pp. 197-205.

Ho, L.C. and Hewitt, J.D. (1986) Fruit development. In: Atherton, J.G. and Rudich, J. (eds), *The Tomato Crop.* Chapman and Hall, London, pp. 201-239.

Ho, L.C., Grange, R.I. and Picken, A.J. (1987) An analysis of the accumulation of water and dry matter in tomato fruit. *Plant, Cell and Environment* 10, 157-162.

Ho, L.C. Belda, R., Brown, M., Andrews, J and Adams, P. (1993) Uptake and transport of calcium and the possible causes of blossom-end rot in tomato. *Journal of Experimental Botany* 44, 509-518.

Ho, L.C., Adams, P., Li, X.Z., Shen, H., Andrews, J. and Xu, Z.H. (1995) Responses of Ca-efficient and Ca-inefficient tomato cultivars to salinity in plant growth, calcium accumulation and blossom-end rot. *Journal of Horticultural Science* 70, 909-918.

Hobson, G.E. and Davies, J.N. (1971) The tomato. In: Hulme, A.C. (ed.) *The Biochemistry of Fruits and Their Products.* Academic Press, London, pp. 437-482.

Massey, D.M. and Winsor, G.M. (1980) Some responses of tomatoes to nitrogen in recirculating solution. *Acta Horticulturae* 98, 127-137.

Massey, D.M., Hayward, A.C. and Winsor, G.W. (1984) Some responses of tomatoes to salinity in nutrient-film culture. *GCRI for 1983 Annual Report,* pp.60-62.

Minamide, R.T. and Ho, L.C. (1993) Deposition of calcium compounds in tomato fruit in relation to calcium transport. *Journal of Horticultural Science* 68, 755-762.

Mitchell, J.P., Shennan, C. and Grattan, S.R. (1991) Developmental changes in tomato fruit composition in response to water deficit and salinity. *Physiologia Plantarum* 83, 177-185.

Nukaya, A., Goto, K., Jang, H., Kano, A. and Ohkawa, K. (1995) Effects of NH_4-N level in the nutrient solution on the incidence of blossom-end rot and gold specks on tomato fruit grown in rockwool. *Acta Horticulturae* 401, 381-388.

Pearce, B.D., Grange, R.I. and Hardwick, K. (1993) The growth of young tomato fruit. I. Effects of temperature and irradiance on fruit grown in controlled environments. *Journal of Horticultural Science* 68, 1-11.

Sheer, C.B. (1975) Calcium related disorder of fruit and vegetables. *HortScience* 10, 361-365.

Sonneveld, C. and Voogt, W. (1990) Response of tomatoes (*Lycopersicon esculentum*) to an unequal distribution of nutrients in the root environment. *Plant and Soil* 124, 251-256.

Tachibana, S. (1988) The influence of withholding oxygen supply to roots by day and night on the blossom-end rot of tomatoes in water culture. *Soilless Culture* 4, 41-50.

van de Vooren, J., Welles, G.W.H. and Hayman, G. (1986) Glasshouse crop production. In: Atherton, J.G. and Rudich, J (eds), *The Tomato Crop.* Chapman and Hall, London, pp. 581-623.

Winsor, G.W. and Adams, P. (1982) The composition and quality of UK and imported tomatoes. *GCRI Annual Report for 1981.* GCRI, Wellesbourne, UK, pp. 154-159.

3 GCRI/Bewley Lecture: Applications of Molecular Biology and Genetic Manipulation to Understand and Improve Quality of Fruits and Vegetables

D. GRIERSON

BBSRC Research Group in Plant Gene Regulation, Plant Science Division, School of Biological Sciences, University of Nottingham, Sutton Bonington Campus, Loughborough, Leics LE12 5RD, UK

I first want to say what a great pleasure and honour it was to be invited to give the GCRI/Bewley Lecture. While walking round the grounds at HRI Wellesbourne, I came across a ginkgo tree planted in the same year that I began my ripening research. This reminds me to acknowledge all those colleagues who have helped me in my research over the years, to whom I am greatly indebted. Ripening and senescence are key processes affecting the quality, appeal, nutritional value and storage life of fruits and vegetables. We chose to study these processes using the tomato as a model ripening system (Grierson *et al.*, 1986a), while realizing that the research would be likely to have implications for other crops. The tomato has proved to be an excellent scientific model, providing the means to identify and study the function of a wide range of 'ripening genes'. It is also an important crop in its own right, with the fresh and processed markets being valued at $3 billion and $13 billion respectively. Within 10 years of the initial identification of important ripening genes, their structure and function have been understood, methods for genetic modification developed, and novel commercial products brought to the market place. In this chapter, I shall discuss what seem to me to have been some of the highlights of tomato ripening research and review how ideas developed from studies on the tomato can be applied to other fruits, vegetables and flowers.

The 1970s and early 1980s were largely taken up with physiological and biochemical studies of ripening, trying to develop a general framework for ideas about how the process worked. This period also saw the introduction of molecular biology techniques, such as gene cloning and plant transformation with *Agrobacterium* vectors, that were to prove so crucial to more recent experimental work. The existence of ripening mutants, some with profound effects, argued for a central role of a genetic programme in ripening, a concept

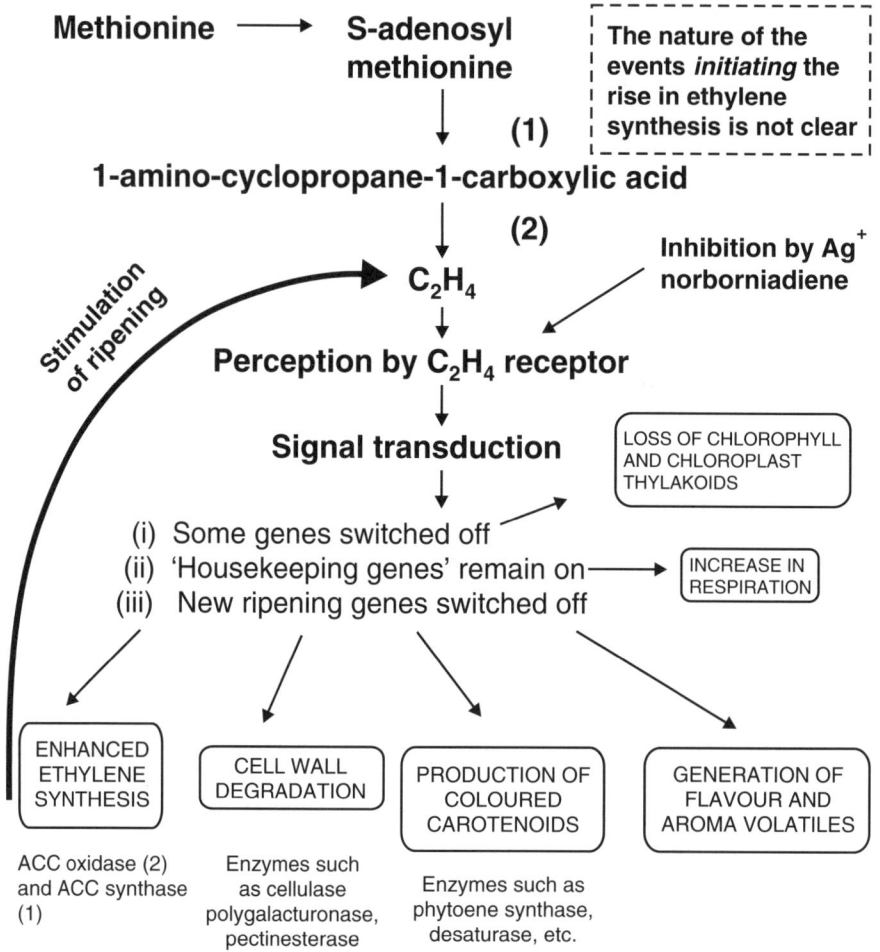

(Synthesis of some mRNAs and proteins occurs independently of C_2H_4)

Fig. 3.1. Ethylene synthesis and the control of ripening in climacteric fruits.

that now seems so obvious; however, in the early 1970s, some authorities certainly needed convincing of this. The general idea is that for climacteric fruits, such as apples, pears, bananas, tomatoes, etc., which show a rise in respiratory output during ripening, ethylene functions as a ripening hormone, which both initiates the onset and accelerates the rate of ripening. The ripening process itself involves a series of downstream events, leading to expression of genes encoding enzymes that bring about changes in colour, flavour, texture and composition. An important feature is that the downstream events can be represented as a branched pathway, because mutation and genetic modification can affect one aspect of ripening, without any apparent effect on another (Fig.

3.1). Biochemical differences between fruits can be explained because there are several ways in which changes in colour, flavour, texture, etc. can be achieved, and different genes have been recruited to the ripening programme in different species during the course of evolution. Changes in gene expression have also been shown to be involved in ripening of non-climacteric fruits, so here we apparently need to identify a different signal, other than ethylene, for ripening initiation.

The history of the cloning of tomato ripening genes has been reviewed elsewhere (Gray *et al.*, 1992). Between 20 and 30 genes have been identified so far, depending on whether one includes genes associated with, but not exclusive to ripening, but this is not nearly enough to account for the complexity of ripening changes known to occur in tomato. There are examples of genes which appear to be expressed specifically during ripening. Other genes, or members of gene families, are involved in ripening but they are also expressed in many cell types and other stages of plant development. Only those genes that have been shown to be involved critically in ripening changes or have featured in transgenic plant experiments will be discussed here. The initial focus was to identify genes whose expression increased at the onset of ripening, and to determine their function. The first ripening cDNA to be identified, originally called TOM6, was shown, in 1986 by cloning and sequencing, to encode the cell wall-metabolizing enzyme polygalacturonase (PG) (DellaPenna *et al.*, 1986; Grierson *et al.*, 1986b). A number of laboratories contributed to studies of this much misunderstood enzyme which catalyses the hydrolysis of polygalacturonic acid chains in unmethylated regions of pectin. There appears to be one gene (Bird *et al.*, 1988) for the endopolygalacturonase that is synthesized *de novo* during tomato ripening (Tucker and Grierson, 1982) but three different polypeptide forms of PG are found in tomato fruit. Early experiments showed that these different isoforms have structural similarities and it was possible to convert one form into another *in vitro*, using a factor present in fruit extracts (Tucker *et al.*, 1981) that was sometimes referred to as the PG converter. This factor has beencloned and is now called the ß-subunit (Zheng *et al.*, 1998). The PG polypeptide can exist in two slightly different forms (PG2A and 2B), possibly differing in carbohydrate content, or can complex with the ß-subunit to produce a third form called PG-1. The three forms of PG are synthesized *de novo* during ripening as a result of transcriptional activation of the PG gene. The inhibition of PG gene expression was first achieved in transgenic plants using antisense genes driven by a constitutive promoter (Smith *et al.*, 1988). Shortly after, the results of very similar experiments were published by scientists from the Calgene Company (Sheehy *et al.*, 1988). PG antisense plants contained reduced levels of PG mRNA and enzyme activity, due to post-transcriptional degradation of PG mRNA in cells in which the endogenous gene and the antisense gene were both transcribed (i.e. ripening fruit). Different transgenic plants had different degrees of down-regulation, suggesting that the position of insertion of the transgene, or possibly some other unknown factor, influenced its effectiveness. This gene silencing was stably inherited, and a gene dosage effect was noted (Smith *et al.*, 1990a). It was also clear from these experiments that inhibition of PG activity had no effect on other ripening attributes such as colour change and ethylene synthesis (Sheehy *et al.*, 1988; Smith *et al.*, 1988).

Shortly after, it was discovered that sense genes could also be used to achieve the same effect (sense-suppression), using approximately half of the full-length PG cDNA under the control of the cauliflower mosaic virus (CaMV) 35S promoter (Smith *et al.*, 1990b). The phenotype of these low PG fruits was identical to that of the PG antisense fruit.

The cell wall pectin, which normally decreases in molecular weight during ripening, was shown to retain a higher molecular weight during ripening of fruits in which the PG gene had been silenced (Smith *et al.*, 1990a). No significant change in firmness could be detected, at least in some varieties, and this and other observations led various researchers to question whether PG was in fact involved in softening. It is now clear from biochemical studies and experiments with transgenic plants, however, that PG has a distinct effect on the textual properties of tomato. Inhibiting its expression, using sense or antisense genes, is the basis for the 'Flavr Savr' tomato, which was marketed in the USA in 1994, and the genetically modified pureé marketed by Sainsbury and Safeway in the UK in 1996. There are thus two different opportunities for exploiting low PG tomatoes. For the fresh market, low PG makes the fruit less susceptible to cracking, splitting and mechanical damage, and fruit, therefore, can be left to ripen longer on the vine before harvesting. Whilst these same advantages may also be significant for processing varieties, the main benefit is in the longer chain pectin, and perhaps the larger size of cell clumps in pureé, both of which affect viscosity and physical properties. The lack of PG also offers the opportunity to process at lower temperatures, which may also bring benefits. The use of low PG tomatoes in pureé is based on sense-suppression. I am pleased to say that Nottingham University receives a royalty on sales.

Other genes were also quickly discovered, including pectinesterase (also involved in cell wall metabolism), phytoene synthase (for carotenoid biosynthesis) and 1-aminocyclopropane-1-carboxylate (ACC) oxidase (for production of the hormone ethylene). In addition, mRNAs encoding histidine decarboxylase, a small heat shock protein and a membrane channel have also been shown to increase in abundance during ripening. At present, it is possible to identify at least 12 genes that have a known function in ripening and a further ten or so for which a clear function remains to be established. If one estimates the number of genes encoding constitutive enzymes required to act with those that are induced during ripening, then we can say that there are scores of genes required for the process. When one considers the apparent complexity of the ethylene signal transduction pathway and other regulatory networks involved, it seems likely that this number will increase very significantly over the next few years.

The function of the first plant phytoene synthase cDNA clone (TOM5) to be discovered was actually demonstrated by inhibiting its expression in transgenic tomatoes using an antisense gene (Bird *et al.*, 1991). Transgenic fruit with low *psy*1 were yellow, rather than red, due to the lack of ß-carotene and lycopene. The residual yellow colour that appeared on ripening was shown to be naringenin chalcone, which is a normal constituent of tomatoes generated from the flavonoid pathway, but which usually is masked by the stronger orange and red colours of the carotenoids. Subsequently, it was shown that two naturally occurring yellow mutants of tomato (*yellowflesh* and *r*^y) contain mutated *psy*1 genes and that the mutation could be complemented by overexpression of the wild-type *psy*1 sequence in the sense orientation under control of the CaMV 35S promoter (Fray and Grierson, 1993). Although this restored the normal red colour to the *yellowflesh* fruit, in some transformants the phenomenon of co-suppression was noted. This was manifested as yellow/white patches on leaves, and in some instances unripe fruit appeared white when they would normally be expected to be green. The explanation for this was that the *psy*1 transgene was suppressing a second related gene, called *psy*2, which is responsible for the synthesis of carotenoids in green tissues. Sense-suppression of *psy*2 led to a failure of other green organs, such as leaves, to accumulate carotenoids, which rendered them susceptible to photobleaching (Fray and Grierson, 1993).

In other experiments, *psy*1 has been overexpressed in wild-type tomatoes under control of the 35S promoter. In some transgenic plants, sense-suppression has been observed, and in others a small increase in carotenoids has been found, probably due to the synthesis of carotenoids in regions of the fruit where it does not normally accumulate in large quantities. In general, however, it has not yet proved possible to bring about a major enhancement of carotenoids in tomatoes by overexpression of *psy*1.

In transgenic plants where the *psy*1 transgene is strongly expressed, the plants frequently are reduced in stature. In severe cases, the plants are dwarf. This is due to utilization of geranylgeranyl diphosphate by the phytoene synthase, which prevents the normal production of gibberellin, and perhaps phytol, and other important metabolites (Fray *et al.*, 1995). Reduction in the levels of these important products, particularly the gibberellins, causes dwarfism. This demonstrates the importance of this area of plant metabolism, and the need for a sophisticated approach to genetic modification if an increase in carotenoids is desired.

Two small multigene families encoding ACC synthase (ACS) and ACC oxidase (ACO) control the biosynthesis of ethylene. The ACS genes were identified by the conventional route of enzyme purification and recognition of bacteria harbouring cloned genes using antibodies. Almost complete inhibition of ACS and, therefore, ethylene production using antisense genes inhibited the change in colour and texture of tomatoes. Ripening changes could be restored by adding ethylene to the picked fruit (Oeller *et al.*, 1991). A reinvestigation of the properties of these fruit has indicated, however, that they still generate sufficient ethylene to induce PG gene expression (Sitrit and Bennett, 1997). Inhibition of ACS has also been achieved in tomato using sense genes (Howie *et al.*, 1996). The genes for ACO were first identified by inhibiting the expression of a specific cDNA (TOM13) by antisense methods, and showing that ACO activity and ethylene production were inhibited in transgenic tomatoes (Hamilton *et al.*, 1990). This was the first example of gene identification using antisense techniques for gene silencing. Inhibiting ACO by 95% permitted normal development of ripening attributes of fruit attached to the plant, but prevented the extreme softening, cracking and spoilage normally associated with over-ripening (Hamilton *et al.*, 1990; Picton *et al.*, 1993), so that the fruit lasted for several weeks. In addition, the senescence of the leaves was delayed for 7–10 days. Interestingly, if the fruit were picked, at the mature green or early breaker stage, they never ripened fully; adding ethylene externally did stimulate colour development, but over-ripening and deterioration did not occur (Picton et al., 1993). When low ethylene fruit are picked, the internal ethylene may be reduced, by diffusion through the calyx scar, thus slowing ripening even further and inhibiting colour production as well (Klee, 1993). Another way of reducing ethylene production is to overexpress foreign genes for S-adenosylmethionine hydrolase (SAMase) or ACC deaminase, which deplete plant cells of the required precursors. These approaches, which rely on overexpression of bacterial enzymes in fruit, have been developed by the Agritope and Monsanto Companies, respectively (Klee et al., 1991; Klee and Tieman, 1997; Kramer et al., 1997).

The first successful demonstration that the ripening behaviour of a fruit other than tomato could be genetically modified was carried out with melon by Ayub et al. (1996), who inhibited *ACO* gene expression using antisense techniques. Melons of the cantaloupe Charentais type were chosen for these experiments because of their good eating quality but poor storage capability. ACO activity in fruit was virtually undetectable, and ethylene production was reduced below 1% of the control level. These transgenic fruit developed a functional abscission layer but remained on the plant for longer. Pigment production in the flesh followed the normal ripening pattern, but the rind remained green while the controls turned yellow. After 10 days storage at 25°C, the low ethylene fruit were still green and retained their shape, whereas the control fruit had a shrivelled yellow rind, showing signs of fungal infection, with soft flesh and a squashed shape. Supplying ethylene to the transgenic melons restored the yellowing phenotype to the rind.

The amino acid sequence predicted from the *ACO* gene sequence clarified thoughts about the enzyme, and enabled it to be solubilized and characterized for the first time (John, 1991). The current view is that the enzyme is soluble in the cytosol, rather than membrane bound as originally thought. It requires iron and ascorbate and is a dioxygenase. Careful studies on expression of the three members of the *ACO* gene family showed that there was a rapid and very substantial increase in gene expression in many situations involving ethylene synthesis (Barry et al., 1996; Blume and Grierson, 1997). This led to the need to revise the view that ACS was the determining step in the ethylene biosynthesis pathway (Grierson et al., 1992) despite spectacular experiments showing that inhibiting ACS activity could almost abolish ripening. Pioneering experiments on ethylene perception and signal transduction in *Arabidopsis* had a dramatic effect on fruit research, and we now know (H.J. Klee, this volume) that the tomato *Never-ripe* mutation affects the ethylene receptor. This particular ethylene receptor is not constitutive, but shows major increases in expression during ripening, senescence, and abscission (Payton et al., 1996). Experiments in progress are uncovering other elements in the signalling chain and may help to clarify the role of ethylene in stimulating expression of ripening genes.

We have used the tomato fruit ethylene receptor cloned by Payton et al. (1996) as a bait to screen a tomato cDNA library in the yeast two-hybrid system.

This has led to the identification of several novel cDNAs encoding putative receptor-interacting proteins. The function of these proteins is being tested by inhibiting their expression *in planta*, using antisense genes.

This research has shown that plant molecular biology and gene technology can greatly improve our understanding of how plants function. This improved insight often suggests ways of improving plants, to make them more suitable for our needs. It is possible to achieve much by molecular breeding that hitherto was thought impossible. This enhanced capability can offer advantages to growers, retailers and consumers, and generate new opportunities for enhancing the quality and nutritional value of food plants.

References

Ayub, R., Guis, M., Ben Amor, M., Gillot, L., Ronstan, J.-P., Latché, A., Bouzayen, M. and Pech, J.-C. (1996) Expression of ACC oxidase antisense inhibits ripening of cantaloupe melon fruits. *Nature Biotechnology* 14, 860–864.

Barry, C.S., Blume, B., Bouzayen, M., Cooper, W., Hamilton, A.J. and Grierson, D. (1996) Differential expression of the 1-aminocyclopropane-1-carboxylate oxidase gene family of tomato. *The Plant Journal* 9, 525–535.

Bird, C.R. Smith, C.J.S., Ray, J.A., Moureau, P., Bevan, M.W., Bird, A.S., Hughes, S., Morris, P.C., Grierson, D. and Schuch, W. (1988) The tomato polygalacturonase gene and ripening specific expression in transgenic plants. *Plant Molecular Biology* 11, 651–662.

Bird, C.R., Ray, J.A., Fletcher, J.D., Boniwell, J.M., Bird, A.S., Teulieres, C., Blain, I., Bramley, P.M. and Schuch, W. (1991) Using antisense RNA to study gene function: inhibition of carotenoid biosynthesis in transgenic tomatoes. *Bio-Technology* 9, 635-639.

Blume, B. and Grierson, D. (1997) Expression of ACC oxidase promoter–GUS fusions in tomato and *Nicotiana plumbaginifolia* regulated by developmental and environmental stimuli. *The Plant Journal* 12, 731–746.

DellaPenna, D., Alexander, D.C. and Bennett, A.B. (1986) Molecular cloning of tomato fruit polygalacturonase: analysis of polygalacturonase levels during ripening. *Proceedings of the National Academy of Sciences* (USA) 83, 6420–6424.

Fray, R.G. and Grierson, D. (1993) Identification and genetic analysis of normal and mutant phytoene synthase genes of tomato by sequencing, complementation and co-suppression. *Plant Molecular Biology* 22, 589–602.

Fray, R.G., Wallace, A., Fraser, P.D., Valero, D., Heddon, P., Bramley, P. and Grierson, D. (1995) Constitutive expression of a fruit phytoene synthase gene in transgenic tomatoes causes dwarfism by redirecting metabolites from the gibberellin pathway. *The Plant Journal* 8, 693–701.

Gray, J., Picton, S., Shabbeer, J., Schuch, W. and Grierson, D. (1992) Molecular biology of fruit ripening and its manipulation with antisense genes. *Plant Molecular Biology* 19, 69–87.

Grierson, D., Hamilton, A.J., Bouzayen, M., Köck, M., Lycett, G.W. and Barton, S. (1992) Regulation of gene expression, ethylene synthesis and ripening in transgenic tomatoes. In: Wray, J.L. (ed.), *Inducible Plant Proteins*. Cambridge University Press, Cambridge, pp. 155–174.

Grierson, D., Maunders, M.J., Slater, A., Ray, J., Bird, C.R., Schuch, W., Holdsworth, M.J., Tucker, G.A. and Knapp, J.E. (1986a) Gene expression during tomato ripening. *Philosophical Transactions of the Royal Society of London* B314, 399-410.

Grierson, D., Tucker, G.A., Keen, J., Ray, J., Bird, C.R. and Schuch, W. (1986b) Sequencing and identification of a cDNA clone for tomato polygalacturonase. *Nucleic Acids Research* 14, 8595-8603.

Hamilton, A., Lycett, G.W. and Grierson, D. (1990) Antisense gene that inhibits synthesis of the hormone ethylene in transgenic plants. *Nature* 346, 284-287.

Howie, W., Lee, K., Baden, J.C., McGugin, C., Bedbrook, J. and Dunsmuir, P. (1996) In: Lycett, G.W. and Tucker, G.A. (eds) *Mechanisms and Applications of Gene Silencing*. Nottingham University Press, Notttingham.

John, P. (1991) How plant molecular biologists revealed a surprising relationship between two enzymes, which took an enzyme out of a membrane where it was not located, and put it into the soluble phase where it could be studied. *Plant Molecular Biology Reporter* 9, 192-194.

Klee, H.J. and Tieman, D. (1997) Potential applications of controlling ethylene synthesis and perception in transgenic plants. In: Kanellis, A.K., Chang, C., Kende, H. and Grierson, D. (eds), *Biology and Biotechnology of the Plant Hormone Ethylene*. Kluwer Academic Publishers, Dordrecht, pp. 289-297.

Klee, H.J., Hayford, M.B. Kretzner, K.A., Barry, G.F. and Kishore, G.M. (1991) Control of ethylene synthesis by expression of a bacterial enzyme in transgenic tomato plants. *The Plant Cell* 3, 1187-1193.

Klee, M.J. (1993) Ripening physiology of fruit from transgenic tomato plants with reduced ethylene synthesis. *Plant Physiology* 102, 911-916.

Kramer, M.G., Kellogg, J., Wagoner, W., Matumura, W., Good, X., Peters, S., Clough, G. and Bestwick, R.K. (1997) Reduced ethylene synthesis and ripening control in tomatoes expressing S-adenosyl-methionine hydrolase. In: Kanellis, A.K., Chang, C., Kende, H. and Grierson, D. (eds), *Biology and Biotechnology of the Plant Hormone Ethylene*. Kluwer Academic Publishers, Dordrecht, pp. 307-319.

Oeller, P.W., Wong, L.M., Taylor, L.P., Pike, D.A. and Theologis, A. (1991) Reversible inhibition of tomato fruit senescence by antisense 1-aminocyclopropane-1-carboxylate synthase. *Science* 254, 427-439.

Payton, S., Fray, R.G., Brown, S. and Grierson, D. (1996) Ethylene receptor expression is regulated during fruit ripening, flower senescence and abscission. *Plant Molecular Biology* 31, 1227-1231.

Picton, S., Barton, S.L., Bouzayen, M., Hamilton, A.J. and Grierson, D. (1993) Altered fruit ripening and leaf senescence in tomatoes expressing an antisense ethylene forming enzyme transgenic. *The Plant Journal* 3, 469-481.

Sheehy, R.E., Kramer, M. and Hiatt, W.R. (1988) Reduction of polygalacturonase activity in tomato fruit by antisense RNA. *Proceedings of the National Academy of Sciences USA* 85, 8805-8809.

Sitrit, Y. and Bennett, A.B. (1997) Regulation of tomato fruit polygalacturonase mRNA accumulation by ethylene: a re-examination. *Plant Physiology* 116, 1145-1150.

Smith, C.J.S., Watson, C., Ray, J., Bird, C.R., Morris, P.C., Schuch, W. and Grierson, D. (1988) Antisense RNA inhibition of polygalacturonase gene expression in transgenic tomatoes. *Nature* 334, 724-726.

Smith, C.J.S., Watson, C.F., Morris, P.C., Bird, C.R., Seymour, G.B., Gray, J.E., Arnold, C., Tucker, G.A., Schuch, W., Harding, S. and Grierson, D. (1990a) Inheritance and effect on ripening of antisense polygalacturonase genes in transgenic tomatoes. *Plant Molecular Biology* 14, 369-379.

Smith, C.J.S., Watson, C.F., Bird, C.R., Ray, J., Schuch, W. and Grierson, D. (1990b)

Expression of a truncated tomato polygalacturonase gene inhibits expression of the endogenous gene in transgenic plants. *Molecular and General Genetics* 224, 477-481.

Tucker, G.A. and Grierson, D. (1982) Synthesis of polygalacturonase during tomato fruit ripening. *Planta* 155, 64-67.

Tucker, G.A., Robertson, N.G. and Grierson, D. (1981) The conversion of tomato fruit polygalacturonase isoenzyme 2 into isoenzyme 1 *in vitro*. *European Journal of Biochemistry* 115, 87-90.

Zheng, L., Heupel, R.C. and DellaPenna, D. (1998) The ß subunit of tomato fruit polygalacturonase isoenzyme 1: isolation, characterization, and identification of unique structural features. *The Plant Cell* 116, 1145-1150.

4 Gene Expression in Ripening Bananas

R. Drury[1]*, C.R. Bird[2] and G.B. Seymour[1]

[1]*Horticulture Research International, Wellesbourne, Warwick CV35 9EF, UK; and* [2]*Zeneca Plant Science, Jealott's Hill Research Station, Bracknell, Berks RG12 6EY, UK*

Introduction

Bananas (*Musa* spp.) are a globally important food crop. It has been estimated that they are the staple food for nearly 400 million people worldwide (Sági *et al.*, 1995), with an estimated 200–300 different clones, including dessert, cooking types and plantains. World production in 1996 was around 85 Mt (FAO, 1997). Despite their importance, conventional breeding has had little impact on improving the agronomic characteristics of the crop. This is because bananas are generally triploid and parthenocarpic, with only a few cultivars having the right combination of postharvest characteristics for international trade. An enormous pool of genetic resources for enhancing the postharvest characteristics of the fruit has, therefore, remained untapped.

Bananas are climacteric fruit where ripening is regulated by ethylene and numerous biochemical changes occur in the ripening fruits including starch metabolism, cell wall disassembly, synthesis of volatile compounds and changes in phenolics. Postharvest characteristics which could be enhanced include better control of green life and shelf life, fruit sweetness and texture. Genetic modification of bananas (May *et al.*, 1995; Sági *et al.*, 1995) offers the opportunity to add desirable characteristics from otherwise inferior types, into a high yielding disease-resistant background. To identify the appropriate target genes, we need to understand the key biochemical and molecular events involved in banana ripening.

The physiology of the banana fruit has been studied intensively over the last 70 years (Marriott, 1980), and this has led to the establishment of recommendations for ripening and storage. Investigations on the nature of the biochemical events involved in ripening have, however, been hampered by the high levels of

---

* Née Rosybel Medina-Suárez. Present address: ICRF Molecular Oncology Unit, Department of Cancer Medicine, ICSM at Hammersmith, London W12 0NN, UK.

starch and phenolics in the fruits. The advent of molecular tools now permits the direct measurement of ripening-related gene expression and this chapter discusses recent work on the isolation and characterization of a wide range of ripening-related genes from banana fruit pulp (Clendennen and May, 1997; Medina-Suárez et al., 1997).

Ripening-related Genes Isolated from Banana Pulp

Genes encoding proteins involved in some of the key postharvest changes are likely to be up-regulated during ripening in the fruit pulp, and, therefore, a library was generated from this tissue and differentially screened. A wide range of ripening-related genes was isolated and their putative identities and patterns of expression are summarized in Table 4.1. These cDNA clones can be grouped into several broad categories based on their putative functions:

- ethylene biosynthesis and respiration,
- carbohydrate metabolism,
- cell wall degradation,
- other ripening-related events.

Ethylene biosynthesis and respiration

1-Aminocyclopropane-1-carboxylate (ACC) oxidase (ACO), which catalyses the conversion of ACC to ethylene, was up-regulated in ripening banana pulp. ACO activity has been shown to increase in pre-climacteric bananas in response to ethylene, prior to the induction of ACC synthase (Liu et al., 1985). López-Gómez et al. (1997) observed that during natural banana ripening an increase in ethylene production was associated with increased ACO activity and transcript levels. When they examined the ACO message in the peel and the pulp, the message was detectable in the pulp at all times, although with reduced intensity in the unripe tissue, but the message was detectable in the peel only in climacteric and post-climacteric fruits (Huang et al., 1997; López-Gómez et al., 1997). The appearance of the ACO message in the pulp earlier than in the peel supports the suggestion that during ripening ethylene production is initiated in the pulp tissue (Domínguez and Vendrell, 1993; Tang et al., 1994; López-Gómez et al., 1997).

An increase in aconitase gene expression in ripening bananas may reflect the increased flux of carbon through the glycolytic pathway as a result of the climateric rise in respiration (Young et al., 1974). Aconitase is present in the cytosol and in the mitochondria (De Bellis et al., 1993, 1994). The mitochondrial aconitase is involved in the citric acid cycle and catalyses the reversible isomerization of citrate to isocitrate via cis-aconitate (Peyret et al., 1995). The cytoplasmic aconitase participates in the glyoxylate cycle together with four glyoxysomal enzymes, playing a role in gluconeogenesis from the stored oil (Hayashi et al., 1995).

Table 4.1. Characterization of up- (pBAN UU) and down-regulated (pBAN UD) cDNA clones from ripening banana pulp.

Clone	Putative identity	Related sequence and accession number[a]	% identity	Base pair overlap
pBAN UU10	ACC oxidase	Apple M81794	68	794
pBAN UU21	Transcriptional activator	Maize L13454	54	311
pBAN UU32	ß-Amylase	Sweet potato D01022	54	789
pBAN UU40	No significant homology			
pBAN UU43	2A11-related clone	Tomato X13743	54	396
pBAN UU55	O-methyltransferase	Maize L14063	60	223
pBAN UU70	Root-related protein	Rice L27208	64	420
pBAN UU80	No significant homology			
pBAN UU84	Pectate lyase	Tomato X55193	66	736
pBAN UU90	Glutamate decarboxylase	Petunia L16797	73	735
pBAN UU91	Aconitase	Pumpkin D29629	76	766
pBAN UU93	Heat shock protein	Tomato X54030	76	771
pBAN UU96	Expressed sequence tag	*Arabidopsis* H36910	69	452
pBAN UU103	Cell wall invertase	Maize U17695	66	567
pBAN UU104	Isoflavonoid reductase	*Arabidopsis* Z49777	61	696
pBAN UU116	Polyubiquitin	Rice X76064	82	625
pBAN UU129	ß-Glucosidase	Barley L41869[b]	60	575
pBAN UU130	No significant homology			
pBAN UU131	S-adenosylhomocysteine hydrolase	Wheat L11872	80	699
pBAN UU136	ß-1,3-Glucanase	Barley M91814	60	800
pBAN UD39	Antifungal protein	Maize U06831	69	517
pBAN UD66	Granule-bound starch synthase	Cassava X74160	64	669
pBAN UD75	Wali 7	Wheat L28008	74	703
pBAN UD93	Chitinase	Cowpea X88801	62	701
pBAN UD120	Extensin	Almond X65718	57	576

[a]Most similar sequence as identified by FASTA search.
[b]Compared with gene sequence edited to remove introns.

Carbohydrate metabolism

Granule-bound starch synthase is one of the key enzymes in the biosynthesis of starch, utilizing ADP-glucose as a substrate. The level of message encoding this enzyme decreases during ripening, and this observation is consistent with a decrease in starch synthesis during this period (Cordenusi and Lajolo, 1995).

ß-Amylase expression was up-regulated in ripening pulp. This enzyme hydrolyses starch at the non-reducing end of the chain, releasing maltose (Smith, 1993). Garcia and Lajolo (1988) observed that in bananas, ß-amylase activity was low in green fruit, but increased significantly (3.5-fold) and simultaneously with the onset of starch hydrolysis, which started during the pre-climacteric phase. They also observed that the activity of ß-amylase increased before that of other hydrolytic enzymes. The observed increase in ß-amylase mRNA levels in ripening bananas is consistent with the biochemical data and supports a role for *de novo* synthesis of this enzyme during banana starch degradation. In other plant tissues, ß-amylase seems to play a key role in starch metabolism. Work on potato leaves and tubers has demonstrated that the major amylolytic activity in potato tissues is due to ß-amylase (Ross and Davies, 1987; Viksø-Nielsen *et al.*, 1997). Also, in *Arabidopsis*, ß-amylase constitutes 80% of the total amylolytic activity in leaves (Lin *et al.*, 1988). It has been shown that the structures of the starch grains in banana fruit are similar to those observed in potato, cereals and mung bean (Kayisu and Hood, 1981; Kayisu *et al.*, 1981) and, therefore, starch metabolism in banana may utilize the same enzymes as in these other tissues (Tucker and Grierson, 1987). There are, however, still fundamental questions to be answered in relation to possible differences in the requirements and conditions for starch degradation in vegetative and storage organs (Viksø-Nielsen *et al.*, 1997). Indeed, the relative importance of hydrolytic and phosphorolytic breakdown of starch in bananas needs to be established.

A cDNA clone encoding a cell wall invertase was isolated from the pulp library (Table 4.1). Invertases catalyse the irreversible cleavage of sucrose into glucose and fructose and can appear in different isoforms inside the same tissue: a non-glycosylated form, characterized by an alkaline pH optimum, is active in the cytosol (Stommel and Simon, 1990), whereas the highly glycosylated acid invertases with a pH optimum of approximately 4.5 occur either in soluble form inside the vacuole or tightly bound to the cell wall (Sturm and Chrispeels, 1990). The exact role of the wall invertase in banana is unknown, but presumably it is involved in the flux of carbon from starch to sugars. The synthesis of sucrose is likely to involve sucrose phosphate synthase (SPS). Very recently, Nascimento *et al.* (1997) have reported the cloning of banana SPS. They found an increase in the level of SPS expression during ripening, along with enhanced enzyme activity, indicating a role for ripening-related synthesis of the enzyme.

Cell wall degradation

Pectate lyase-like sequences were the most abundant ripening-related clones identified from the banana pulp library. Indeed, there appear to be two different pectate lyase-like genes expressed in banana fruit each with slightly different patterns of expression (Medina-Suárez, 1998). Pectate lyases are found commonly as pathogen-secreted enzymes, that macerate plant tissues by random cleavage of ß-1,4-linked galacturonosyl residues of cell wall pectins (Collmer and Keen, 1986). Enzymatic cleavage of glycosidic bonds occurs through a ß-elimination, which produces an unsaturated C4–C5 bond in the galacturonosyl moiety at the non-reducing end of the polysaccharide and generates 4,5-unsaturated oligogalacturonates (Yoder *et al.*, 1993). In pathogens, there are multiple and independently regulated pectate lyase isoenzymes which are encoded by different genes (Collmer and Keen, 1986; Kelemu and Collmer, 1993; Alfano *et al.*, 1995).

Several plant sequences with homology to pectate lyases have been reported, including those from pollen grains of tomato (Wing *et al.*, 1989), short ragweed (Rafnar *et al.*, 1991), tobacco (Rogers *et al.*, 1992) and maize (Turcich *et al.*, 1993), and from stylar tissue in tomato and tobacco (Budelier *et al.*, 1990).

Very recently, pectate lyase cDNAs from fruits have been reported for the first time, e.g. strawberry (Medina-Escobar *et al.*, 1997) and banana (Domínguez-Puigjaner *et al.*, 1997; Medina-Suárez *et al.*, 1997). The substrate specificity of the published clones from the majority of other plant tissues has not been confirmed, although Medina-Escobar *et al.* (1997) demonstrated pectate lyase activity for their strawberry cDNA clone by overexpression in yeast. The pectate lyase-like sequences in banana may encode enzymes involved in cell wall degradation and softening.

At least two extensin genes are expressed during banana ripening, one showing constitutive expression (Domínguez-Puigjaner *et al.*, 1997) and the other being down-regulated in ripening banana pulp (Medina-Suárez *et al.*, 1997). Extensins are cell wall structural glycoproteins rich in hydroxyproline. They are induced by environmental stresses such as mechanical wounding and pathogen attack, and may also play a role in cell wall strengthening during development. In tobacco, *in situ* hybridization experiments have shown that accumulation of extensin mRNA occurs at certain stages of development in cells which require reinforcement of their cell walls (Tiré *et al.*, 1994). Also, extensin genes have been shown to be regulated by the tensile stress experienced by the cortical cells, so that as the level of stress increases, so does the intensity of expression (Shirsat *et al.*, 1996). Down-regulation of these proteins in banana pulp may, therefore, reflect biochemical events designed to weaken or change the structure of the pulp cell walls (Medina-Suárez *et al.*, 1997).

The banana clone pBANUU129 (Table 4.1) shares sequence homology with a barley ß-glucosidase cDNA. This barley clone belongs to Family I of glycosyl hydrolases and appears to play a role in degrading cell wall polysaccharides of the endosperm during germination (Leah *et al.*, 1995). High molecular weight ß-linked polysaccharides are degraded by hydrolases during germination and are thought to be metabolized further by ß-glucosidase (Leah *et al.*, 1995). The function of the protein encoded by the banana ß-glucosidase-like sequence is unknown. Perhaps its role is analogous to that in developing seeds, facilitating starch mobilization by cell wall disassembly in the pulp.

ß-1,3-Glucanases were found to be up-regulated in ripening banana pulp. These enzymes hydrolyse ß-linked glucans, and are believed to be involved in germination, cell growth, plant defence against pathogens, and flowering, and they also have been implicated in fruit and seed maturation, abscission and senescence, and tissue differentiation (Simmons, 1994). ß-1,3-Glucanases have been found to increase during ripening in a number of fruits, and it has been suggested that they may be involved in cell wall degradation leading to fruit softening (Hinton and Pressey, 1980). The role of the enzyme in ripening bananas is unknown.

Other ripening-related events

Both Medina-Suárez *et al.* (1997) (see Table 4.1) and Clendennen and May (1997) isolated chitinase (class III) clones from their pulp libraries. Chitinases are a group of pathogenesis-related clones which have a significant role in defence against invading fungal pathogens. They hydrolyse chitin, which is the major component of the exoskeleton of insects and of the cell wall of most fungi (Collinge *et al.*, 1993), and the enzyme has been shown to exhibit antifungal activity *in vitro* (Mauch *et al.*, 1988; Arlorio *et al.*, 1992). It has been suggested, however, that the effectiveness of chitinases against fungi may depend on the simultaneous action of ß-1,3-glucanase or other antifungal substances (Collinge *et al.*, 1993). The role of these enzymes in bananas requires further investigation. Two clones with homology to glutamate decarboxylase (GAD) were isolated from the banana pulp cDNA library (Medina-Suárez *et al.*, 1997). GAD catalyses the conversion of glutamic acid to γ-aminobutyric acid (GABA), and in vertebrates and invertebrates, GABA functions as a major neurotransmitter modulating the function of ion channels (Baum *et al.*, 1993, and references therein). GAD has been shown to be up-regulated in ripening tomatoes (Gallego *et al.*, 1995), but its role in fruit ripening is unknown. An isoflavonoid reductase (IFR)-like message was present in unripe pulp and was up-regulated after ethylene treatment. Recent findings indicate that an IFR-like sequence from *Arabidopsis* can confer resistance to oxidative stress in yeast (Babiychuk *et al.*, 1995). The role of IFR in banana ripening is unclear, but its expression may be related to an increase in the oxidative environment during ripening. Other clones of unknown functions included a sequence related to the 2A11 fruit-specific cDNA from tomato, a clone with some homology to nodulin (*Nms-25*, Végh *et al.*, 1990), and a sequence with homology to root-specific proteins (John *et al.*, 1992). Although the biochemical function of many of these cDNA clones remains unknown, the rapid rate at which plant genes are being isolated suggests it will soon be possible to assign these sequences a putative function by database comparisons.

Conclusions

Clendennen and May (1997) noted that although banana ripening is induced by ethylene, this hormone is a signal for other physiological changes including senescence. Therefore, some clones isolated by us and by the above authors may be regulated by ethylene, but may not be involved directly in the ripening process, e.g. pathogenesis-related proteins such as glucanase, chitinase and thaumatin. However, the banana clones described so far provide a starting point for understanding the biochemical and molecular basis of ripening in this fruit. The next step will be to examine the role of the various gene sequences we have isolated in key postharvest events by using transgenic bananas.

Acknowledgements

This research was supported by Instituto Nacional de Investigaciones Agrarias Ministerio de Agricultura, Español, Zeneca Plant Science and the Biotechnology and Biological Sciences Research Council, UK.

References

Alfano, J.R., Ham, J.H. and Collmer, A. (1995) Use of Tn5tac 1 to clone a *pel* gene encoding a highly alkaline asparagine-rich pectate lyase isozyme from an *Erwinia chrysantemi* EC16 mutant with deletions affecting the major pectate lyase isozymes. *Journal of Bacteriology* 177, 4553-4556.

Arlorio, M., Ludwig, A., Boller, T. and Bonfante, P. (1992) Inhibition of fungal growth by plant chitinases and ß-1,3-glucanase. A morphological study. *Protoplasma* 171, 34-43.

Babiychuk, E., Kushnir, S., Belles-Boix, E., Van Montagu, M. and Inzé, D. (1995) *Arabidopsis thaliana* NADPH oxidoreductase homologs confer tolerance of yeasts toward the thiol-oxidizing drug Diamide. *Journal of Biological Chemistry* 270, 26224-26231.

Baum, G., Chen, K., Arazi, T., Takatsuji, H. and Fromm, H. (1993) A plant glutamate decarboxylase containing a calmodulin binding domain. *Journal of Biological Chemistry* 268, 19160-19167.

Budelier, K.A., Smith, A.G. and Gasser, G.S. (1990) Regulation of a stylar transmiting tissue-specific gene in wild-type and transgenic tomato and tobacco. *Molecular and General Genetics* 224, 183-192.

Clendennen, S.K. and May, G.D. (1997) Differential gene expression in ripening banana (*Musa acuminata* cv. Grand Nain) fruit. *Plant Physiology* 115, 463-469.

Collinge, D.B., Kragh, K.M., Mikkelsen, J.D., Nielsen, K.K., Ramussen, U. and Vad, K. (1993) Plant chitinases. *The Plant Journal* 3, 31-40.

Collmer, A. and Keen, N.T. (1986) The role of pectic enzymes in plant pathogenesis. *Annual Review of Phytopathology* 24, 383-409.

Cordenusi, B.R. and Lajolo, F.M. (1995) Starch breakdown during banana ripening: sucrose synthase and sucrose phosphate synthase. *Journal of Agriculture and Food Chemistry* 43, 347-351.

De Bellis, L., Tsugeki, R., Alpi, A. and Nishimura, M. (1993) Purification and characterisation of aconitase isoforms from etiolated pumpkin cotyledons. *Physiologia Plantarum* 88, 485-492.

De Bellis, L., Hayashi, M., Biagi, P.P., Hara-Nishimura, I., Alpi, A. and Nishimura, M. (1994) Inmunological analysis of aconitase in pumpkin cotyledons: the absence of aconitase in glyoxysomes. *Physiologia Plantarum* 90, 757-762.

Domínguez, M. and Vendrell, M. (1993) Ethylene biosynthesis in banana fruit: Evolution of EFE activity and ACC levels in peel and pulp during ripening. *Journal of Horticultural Science* 68, 63-70.

Domínguez-Puigjaner, E., LLop, I., Vendrell, M. and Prat, S. (1997) A cDNA clone highly expressed in ripe banana fruit shows homology to pectate lyases. *Plant Physiology* 114, 1071-1076.

FAO (1997) *Food and Agriculture Organization of the United Nations, Production Yearbook 1996*, Vol 50. FAO, Rome.

Gallego, P.P., Whotton, L., Picton, S., Grierson, D. and Gray, J.E (1995) A role for glutamate decarboxylase during tomato ripening: the characterisation of a cDNA encoding a putative glutamate decarboxylase with a calmodulin-binding site. *Plant Molecular Biology* 27, 1143-1151.

Garcia, E. and Lajolo, F.M. (1988) Starch transformation during banana ripening: the amylase and glucosidase behavior. *Journal of Food Science* 53, 1181-1186.

Hayashi, M., De Bellis, L., Alpi, A. and Nishimura, M. (1995) Cytosolic aconitase participates in the glyoxylate cycle in ethyolated pumpkin cotyledons. *Plant and Cell Physiology* 36, 669-680.

Hinton, D.M. and Pressey, R. (1980) Glucanases in fruits and vegetables. *Journal of the American Society of Horticultural Science* 105, 499–502.

Huang, P.-L., Do, Y.-Y., Huang, F.-C., Thay, T.-S. and Chang, T.-W. (1997) Characterisation and expression analysis of a banana gene encoding 1-aminocyclopropane-1-carboxylate oxidase. *Biochemistry and Molecular Biology International* 41, 941–950.

John, I., Wang, H., Held, B.M., Wurtele, E.S. and Colbert, J.T. (1992) An mRNA that specifically accumulates in maize roots delineates a novel subset of developing cortical cells. *Plant Molecular Biology* 20, 821–831.

Kayisu, K. and Hood, L.F. (1981) Molecular structure of banana starch. *Journal of Food Science* 46, 1894–1897.

Kayisu, K., Hood, L.F. and Van Soest, P.J. (1981) Characterisation of starch and fiber of banana fruit. *Journal of Food Science* 46, 1885–1890.

Kelemu, S. and Collmer, A. (1993) *Erwinia chysantemi* EC 16 produces a second set of plant inducible pectate lyase isozymes. *Applied and Environmental Microbiology* 59, 1756–1761.

Leah, R., Kigel, J., Svendsen, I.B. and Mundy, J. (1995) Biochemical and molecular characterisation of a barley seed ß-glucosidase. *Journal of Biological Chemistry* 270, 15789–15797.

Lin, T.P., Caspar, T., Somerville, C. and Preiss, J. (1988) Isolation and characterisation of a starchless mutant of *Arabidopsis thaliana* lacking ADP glucose pyrophosphorylase activity. *Plant Physiology* 86, 1131–1135.

Liu, Y., Hoffman, N.E. and Yang, S.-F. (1985) Promotion by ethylene of the capability to convert 1-aminocyclopropane-1-carboxylic acid to ethylene in preclimacteric tomato and cantaloupe fruits. *Plant Physiology* 77, 407–411.

López-Gómez, R., Campbell, A., Dong, J.-G,, Yang, S.-F. and Gómez-Lim, M.A. (1997) Ethylene biosynthesis in banana fruit: isolation of a genomic clone to ACC oxidase and expression studies. *Plant Science* 123, 123–131.

Marriott, J. (1980) Bananas. Physiology and biochemistry of storage and ripening for optimum quality. *CRC Critical Reviews on Food Science and Nutrition* 13, 41–88.

Mauch, F., Mauch-Mani, B. and Boller, T. (1988) Antifungal hydrolases in pea tissue. II. Inhibition of fungal growth by combinations of chitinase and ß-1,3-glucanase. *Plant Physiology* 88, 936–942.

May, G.D., Afza, R., Mason, H.S., Wiecko, A., Novak, F.J. and Arntzen, C.J. (1995) Generation of transgenic banana (*Musa acuminata*) plants via *Agrobacterium*-mediated transformation. *Bio-Technology* 13, 486–492.

Medina-Escobar, N., Cárdenas, J., Moyano, E., Caballero, J.L. and Muñoz-Blanco, J. (1997) Cloning, molecular characterization and expression pattern of a strawberry ripening-specific cDNA with sequence homology to pectate lyase from higher plants. *Plant Molecular Biology* 34, 867–877.

Medina-Suárez, R. (1998) Molecular biology of banana fruit ripening. PhD thesis, Cranfield University.

Medina-Suárez, R., Manning, K., Fletcher, J., Aked, J., Bird, C.R. and Seymour, G.B. (1997) Gene expression in the pulp of ripening bananas. *Plant Physiology* 115, 453–461.

Nascimento, J.R.O., Cordenusi, B.R., Lajolo, F.M. and Alcocer, F.M.C. (1997) Banana sucrose-phosphate synthase gene expression during fruit ripening. *Planta* 203, 283–288.

Peyret, P., Perez, P. and Alric, M. (1995) Structure, genomic organisation, and expression of the *Arabidopsis thaliana* aconitase gene. *Journal of Biological Chemistry* 270, 8131–8137.

Rafnar, T., Griffith, I.J., Kuo, M., Bond, J.F., Rogers, B.L. and Klapper, D.G. (1991) Cloning of Amb a I (antigen E), the major allergen family of short ragweed pollen. *Journal of Biological Chemistry* 226, 1229-1236.

Rogers, H.J., Harvey, A. and Lonsdale, D.M. (1992) Isolation and characterization of a tobacco gene with homology to pectate lyase which is specifically expressed during microsporogenesis. *Plant Molecular Biology* 20, 493-502.

Ross, H.A. and Davies, H.V. (1987) Amylase activity in potato tubers. *Potato Research* 30, 675-678.

Sági, L., Panis, B., DeSmet, K., Remy, S., Swennen, R. and Cammue, B.P.A.(1995) Genetic transformation of banana and plantain (*Musa* spp) via particle bombardment. *Bio-Technology* 13, 481-485.

Shirsat, A.H., Bell, A., Spence, J. and Harris, J.N. (1996) The *Brassica napus extA* extensin gene is expressed in regions of the plant subject to tensile stress. *Planta* 199, 618-624.

Simmons, C.R. (1994) The physiology and molecular biology of plant 1,3-ß-D-glucanases and 1,3;1,4-ß-D-glucanases. *Critical Reviews in Plant Sciences* 13, 325-387.

Smith, C.J. (1993) Carbohydrate chemistry. In: Lea, P.J. and Leegood, R.C. (eds), *Plant Biochemistry and Molecular Biology*. John Wiley and Sons, London, pp. 73-112.

Stommel, J.R. and Simon, P.W. (1990) Multiple forms of invertase from *Daucas carota* cell cultures. *Phytochemistry* 29, 2087-2089.

Sturm, A. and Chrispeels, M.J. (1990) cDNA cloning of carrot extracellular ß-fructosidase and its expression in response to wounding and bacterial infection. *The Plant Cell* 2, 1107-1119.

Tang, X., Gomes, A.M.T.R., Bhatia, A. and Woodson, W.R. (1994) Pistil-specific and ethylene-regulated expression of 1-aminocyclopropane-1-carboxylate oxidase genes in petunia flowers. *The Plant Cell* 6, 1227-1239.

Tiré, C., De Rycke, R., De Loose, M., Inzé. D., Van Montagu, M. and Engler, G. (1994) Extensin gene expression is induced by mechanical stimuli leading to local cell wall strengthening in *Nicotiana plumbaginifolia*. *Planta* 195, 175-181.

Tucker, G.A. and Grierson, D. (1987) Fruit ripening. In: Stumpf, P.K. and Conn, E.E. (eds), *Biochemistry of Plants*. Academic Press, London, pp. 265-381.

Turcich, M.P., Hamilton, D.A. and Mascarenhas, J.P. (1993) Isolation and characterisation of pollen-specific maize genes with sequence homology to ragweed allergens and pectate lyases. *Plant Molecular Biology* 23, 1061-1065.

Végh, Z., Vincze, É., Kadirov, R., Tóth, G. and Kiss, G.B. (1990) The nucleotide sequence of a nodule-specific gene, *Nms-25* of *Medicago sativa*: its primary evolution via exon-shuffling and retrotransposon-mediated DNA arrangements. *Plant Molecular Biology* 15, 295-306.

Viksø-Nielsen, A., Christensen, T.M.I.E., Bojko, M. and Marcussen, J. (1997) Purification and characterisation of ß-amylase from leaves of potato (*Solanum tuberosum*). *Physiologia Plantarum* 99, 190-196.

Wing, R.A., Yamaguchi, J., Larabell, S.K., Ursin, V.M. and McCormick, S. (1989) Molecular and genetic characterisation of two pollen expressed genes that have sequence similarity to pectate lyases from the pollen pathogen *Erwinia*. *Plant Molecular Biology* 14, 17-28.

Yoder, M.D., Keen, N.T. and Jurnak, F. (1993) New domain motif: the structure of pectate lyase C, a plant virulence factor. *Science* 260, 1503-1507.

Young, R.E., Salminen, S. and Sornsrivichai, P. (1974) Enzyme regulation associated with ripening in banana fruit. *Colloques Internationeaux, CNRS 238*, 271-279.

5 Genes for Fruit Quality in Strawberry

K. MANNING

Horticulture Research International, Wellesbourne, Warwick CV35 9EF, UK

Introduction

The strawberry is cultivated in all temperate regions of the world. Around 2.5 Mt are grown annually, the main areas of production being North America, Mexico, Europe (principally Belgium, Holland, Italy, Poland and Spain) and Japan (Manning, 1993). The cultivated strawberry, *Fragaria ananassa*, is an interspecific hybrid octaploid complex derived from two other octaploid species *F. chiloensis* and *F. virginiana*. Modern strawberry cultivars have resulted from hybridizations from within a relatively narrow genetic background, and this has produced inbreeding depression leading to an increased susceptibility of the crop to pests, diseases and environmental stresses (Harrison *et al.*, 1997). There is now considerable interest in broadening the genetic base of this crop by incorporating useful traits from wild species. One of the main limitations to expanding genetic diversity in *Fragaria* is the problem of incompatibility arising from the range of ploidy levels. For example, crosses between the diploid *F. vesca* and *F. ananassa* produce few viable seedlings that are true hybrids.

The possibility of plant genetic modification now offers an alternative means of crop enhancement. Provided that suitable genes can be identified, we can contemplate the improvement of commercially important traits by selectively modifying the expression of their associated genes. A number of traits for possible genetic manipulation have been proposed in strawberry to improve crop production, mechanical harvesting and fruit quality (Jain and Pehu, 1992). This chapter describes how recent progress in unravelling the molecular and biochemical events characterizing the ripening of strawberry is helping to identify potential candidate genes for improving the quality attributes of the fruit.

Early Biochemistry

Our understanding of the biochemical and molecular determinants affecting ripening in the strawberry has lagged behind that of some other fruits, notably tomato. Technical difficulties associated with high levels of phenolic compounds and viscous polysaccharides in the fruit have hindered the purification of proteins and the isolation of nucleic acids from this tissue. Relatively few enzymes have been assayed from strawberry, and only phenylalanine ammonia lyase (PAL) has been purified to homogeneity (Given *et al.*, 1988a). Changes in the activity of enzymes believed to be involved in cell wall metabolism and pigmentation have been reported, but the insights gained from these studies have been limited.

Hormonal Regulation of Ripening

The ripening of strawberry exhibits typical non-climacteric behaviour in that it appears to be totally independent of the hormone ethylene, both for initiation and for maintenance. Removal of the achenes from one half of a strawberry fruit at the mature green stage was shown to accelerate ripening in the corresponding part of the receptacle (Given *et al.*, 1988b). Application of the synthetic auxin analogue 1-naphthaleneacetic acid (NAA) to the de-achened portion of the receptacle inhibited its ripening, whereas the inactive auxin analogue, phenoxyacetic acid (POA), was relatively ineffective. Auxin appears to be involved in all aspects of strawberry fruit development since it is also essential for the growth of the fruit (Nitsch, 1950). The achenes, believed to be the source of auxin in the receptacle, may play a key role in the temporal regulation of ripening. It is thought that the decline in auxin synthesis in the achenes as the fruit matures induces a reduction in the concentration of auxin in the receptacle. When the concentration of auxin reaches a critical level in the receptacle, ripening is no longer suppressed.

Gene Expression and Fruit Development

In order to study gene expression in developing strawberry fruit, an effective method for isolating RNA free of contaminating polyphenols and polysaccharides was developed (Manning, 1991). Changes in gene expression were examined by translating total RNA *in vitro* and separating the translated polypeptides by two-dimensional polyacrylamide gel electrophoresis (Manning, 1994). Changes in the abundance of more than 50 polypeptides were detected between the immature green and over-ripe stages, with most occurring at or just before the onset of anthocyanin accumulation. These changes represented groups of genes whose expression either generally increased or decreased during receptacle development. Other genes were expressed in immature and ripe fruit, but fell silent as or just before the fruit changed colour.

The pattern of gene expression in fruits whose ripening was accelerated by removing the achenes was indistinguishable from that in normally ripened fruit (Manning, 1994). When ripening was inhibited by NAA, gene expression in the de-achened receptacle was similar to that of unripe fruit. The inactive auxin analogue POA neither inhibited ripening nor modified the pattern of gene expression seen in ripe fruit. Thus, auxin negatively regulates a group of ripening-enhanced genes in strawberry fruit.

Cloning of Ripening-related Genes

Using information gained from the translation studies, cDNA libraries were constructed from mRNA isolated from fruit at the white and fully ripe stages. Ripe fruit cDNA was differentially screened with ripe and white cDNA. More than 100 ripening-related clones were isolated from approximately 1300 clones used in the initial screen, representing an abundance of about 6%. The ripening-enhanced clones were sequenced and the sequences compared with known sequences in nucleotide and protein databases. From database homologies, 26 different gene families were identified.

Genes Related to Quality Traits

Table 5.1 shows a selection of ripening-related genes and associated traits related to quality. These are described in more detail below.

Table 5.1. Selected strawberry ripening genes and their putative functions related to quality.

Gene	Function
O-Methyltransferase Chalcone synthase Chalcone reductase Flavanoid-3-hydroxylase UDP–glucosyl transferase UDP–glucuronosyl transferase	Anthocyanin/phenolic
Cellulase	Texture
Acyl carrier protein	Lipid biosynthesis
Sucrose transporter	Sugar accumulation
Cysteine proteinase	Protein turnover
Elongation factor	Metabolic rate
Pyruvate decarboxylase	Respiration/flavour
Protein kinase	Signalling – development

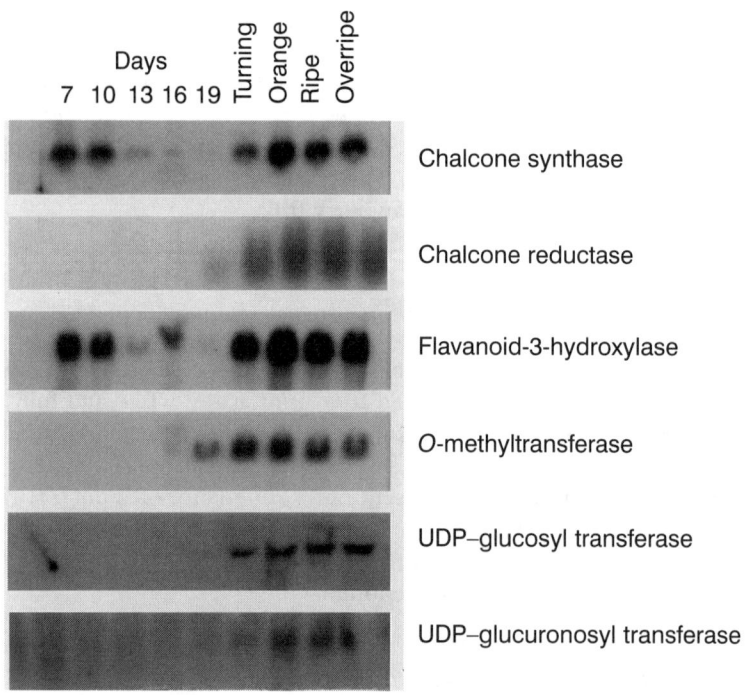

Fig. 5.1. Expression of phenylpropanoid genes in developing strawberry fruit.

Phenylpropanoid metabolism: relationship to colour and flavour

Amongst the ripening-related genes identified in strawberry are six families of structural genes putatively involved in phenylpropanoid metabolism. Northern analysis of their expression during fruit development shows differential patterns of regulation (Fig. 5.1). Flavanoid-3-hydroxylase and chalcone synthase are both expressed in immature fruit, and their transcripts decrease to undetectable levels at the white stage (day 19) before a dramatic increase as anthocyanin pigments are formed in the turning fruit. In contrast, the expression of chalcone reductase, O-methyltransferase, UDP–glucosyl transferase and UDP–glucuronosyl transferase is delayed in developing fruit until the beginning of anthocyanin accumulation.

Phenylpropanoid compounds affect the quality of strawberry fruit in a number of ways. Colour is an important attribute to the consumer when judging fruit ripeness. Pelargonidin-3-glucoside is the principal red pigment in ripe strawberry along with minor amounts of cyanidin-3-glucoside. In flowers, it is known that anthocyanin accumulation requires the coordinated expression of several genes in the biosynthetic pathway. In strawberry, *de novo* synthesis of PAL, a key enzyme in the formation of phenolic compounds, is required for anthocyanin production (Given *et al.*, 1988a). The activity of UDP–glucosyl transferase, the terminal step in the biosynthesis of pelargonidin-3-glucoside, parallels the accumulation of anthocyanin (Given et al., 1988c).

Colour intensity and spectral quality will depend on the concentration and types of pigment present. We have isolated strawberry cDNAs encoding flavanoid-3-hydroxylases from four gene families. These enzymes regulate the interconversions between phenylpropanoid intermediates, the relative expression of their genes affecting the types of anthocyanins produced. In the food industry, strawberry anthocyanins are well known for being unstable during processing, and fruit colour is adversely affected by heat and freezing treatments. Glycosylation of the proanthocyanidins catalysed by UDP–glucosyl and UDP–glucuronosyl transferases has a profound influence on colour stability. Other modifications to the phenylpropanoid ring such as the addition of methyl groups catalysed by the enzyme O-methyltransferase could also affect pigment stability.

Although anthocyanins may be the end-products of phenylpropanoid metabolism in ripening strawberry, the intermediate steps in the pathway may have a significant effect on the composition of other phenols. Chalcone synthase acts at a branch pathway in phenylpropanoid metabolism and is believed to regulate the balance between flavonoids and isoflavonoids when co-acting with chalcone reductase (Colliver et al., 1997). The patterns of expression of chalcone synthase and flavanoid-3-hydroxylase indicate that a switch in phenolic acid metabolism may occur as the strawberry matures. Phenols contribute to the astringent properties of many unripe fruits, and this may be a protective mechanism to ensure that fruits are consumed only when they are ripe and their seeds are fertile. The astringent phenols in immature strawberry are likely to be the precursors of anthocyanins formed in the ripe fruit.

Flavonols: nutritional and medicinal properties

In addition to their aesthetic properties and their impact on flavour, flavonols may provide significant nutritional and medicinal benefits. Epidemiological studies indicate that compounds such as quercetin and kaempferol, which are present in strawberry, have useful antioxidant properties. Other phenolics present in strawberries are claimed to have antiulcerogenic and anticarcinogenic effects.

Cellulase

One of the most important properties affecting strawberry fruit quality is texture. The firmness of the fruit is the major factor determining eating quality, postharvest deterioration and susceptibility to mechanical damage. The mechanism of softening of any fruit is far from certain, and no single enzyme has been proved responsible for ripening-associated changes in texture. In soft fruits such as the strawberry, there are particularly rapid changes in firmness during ripening. Although alterations in the cell wall of strawberry during ripening involve the release of soluble polyuronides, there is little corresponding polygalacturonase activity in the fruit. Similarly, pectin methylesterase activity is not

Table 5.2. Cellulase activities in climacteric and non-climacteric fruits at the ripe stage.

Fruit	Cultivar	Activity g^{-1} fresh weight	Relative activity
Avocado	Hass	6.99×10^{-5}	33,000
Pepper		1.49×10^{-6}	702
Raspberry		5.06×10^{-7}	239
Strawberry	Elsanta	2.12×10^{-7}	100
Apple	Golden Delicious	7.61×10^{-9}	3.6
Tomato	Ailsa Craig	5.93×10^{-9}	2.8

correlated obviously with cell wall hydrolysis. The enzyme cellulase, however, has been detected in ripe and over-ripe strawberry. Different viscometric methods have been used for measuring cellulase activity in fruits, and it is difficult to compare values reported in the literature. We have therefore assayed cellulase activity from a range of fruits using a single viscometric method and found that cellulase activity varied widely between different fruits, with strawberries lying in the middle of the range (Table 5.2).

We have isolated, by differential screening, three cDNA clones from strawberry encoded by the same cellulase gene. The role of cellulase in strawberry fruit softening is being investigated using several approaches. We have purified a protein from ripe strawberry fruit to apparent homogeneity and, from its N-terminal amino acid sequence, have confirmed it to be cellulase.

Although several ripening-related cDNAs encoding cellulase have been characterized from a number of fruits, definitive experiments to test the effects of altered cellulase activity upon fruit softening have not been reported. We are investigating how modifying cellulase activity affects the texture of strawberry fruit by down-regulating cellulase gene expression in transformed strawberry plants. Transformants have been prepared with constructs containing the cellulase sequence in the antisense and sense (for co-suppression) orientation, with plants currently at the flowering stage.

Expansins

We have examined the distribution of firmness in ripe strawberry fruit from an F_1 generation obtained by crossing a soft cultivar with a firm cultivar (Fig. 5.2). The population exhibited a nearly normal distribution of firmness, indicating that texture of the ripe fruit is likely to be a multigene trait involving several types of wall-modifying proteins.

One of these might be the expansins, a class of plant proteins without hydrolytic activity that have the unique property of facilitating the expansion of plant cell walls. This cell wall–loosening action is believed to involve the rearrangement of hydrogen bonds between the cellulose microfibrils and xyloglucans. Originally discovered in hypocotyl tissues, it is now apparent that expansins influence cell wall architecture in all plant tissues, including fruits. Expansins are of particular interest in fruits because they may have a major role

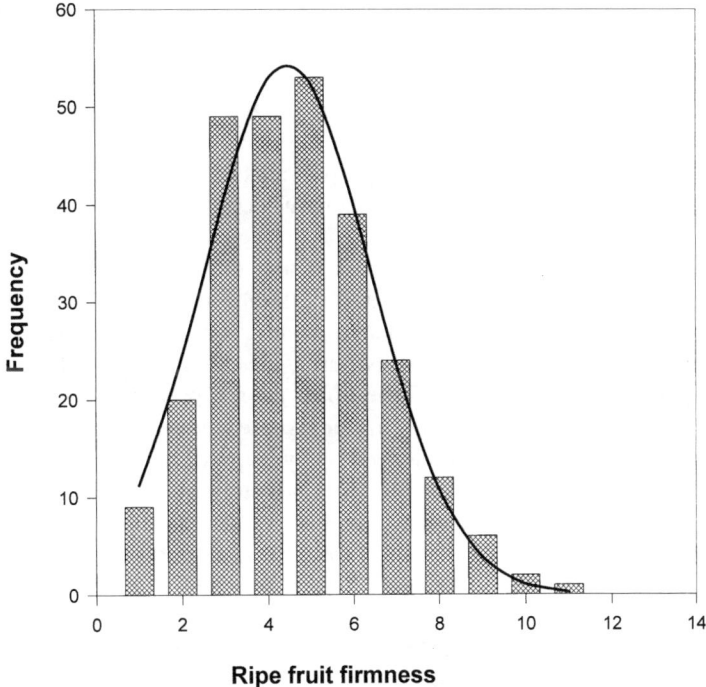

Fig. 5.2. Segregation of the firmness of ripe fruit in an F_1 generation from a cross between a soft and a firm cultivar. Firmness was measured as peak compression force (Newtons).

in fixing the juxtapositions of polysaccharides within the cell wall architecture, thereby influencing their susceptibility to the action of hydrolytic enzymes during ripening. We are examining the role of expansins in relation to changes in texture in strawberry fruit, which continue to grow during the ripening phase. A number of expansin sequences have been cloned from strawberry genomic DNA, consistent with the multigene character of this family of proteins in other species. These clones will be used to probe the expression of the corresponding genes and this information will be used to generate transgenic plants with altered levels of expansins.

Acyl carrier protein

One of the surprising clones to have been isolated from strawberry encodes acyl carrier protein (ACP). As an essential component of the fatty acid biosynthetic machinery in plants, ACPs have been studied mostly in tissues that accumulate lipids.

Northern analysis of developing strawberry fruits shows that the expression of ACP is barely detectable in immature stages, but increases markedly at the onset of ripening, remaining high into the over-ripe stage. Strawberry fruit do not accumulate fatty acids and there are only minor differences in the ratios of the various types of fatty acid found in unripe and ripe fruit. However, it is known that the fruit acquires the ability to synthesize volatile flavour compounds as it ripens. Ripe strawberries have one of the most complex aroma profiles of any fruit, with 281 odour-active compounds being reported (Zabetakis and Holden, 1997). These compounds comprise acids, alcohols, aldehydes, esters, terpenoids and miscellaneous chemicals, including 2,5-dimethyl-4-methoxy-2*H*-furan-3-one, which is derived from 6-deoxyfructose and has a strawberry-like odour. It is apparent that no compound by itself mimics the complete aroma of the fruit, and certain compounds have different flavour characteristics according to their concentrations. Several of the more important odour-active substances in strawberry, for example *trans*-2-hexenal, are probably derived from lipids via the lipoxygenase pathway. Linoleic and linolenic acids, both C_{18} unsaturated fatty acids, are converted to their corresponding hydroperoxides and then cleaved by hydroperoxide lyase to form C_6/C_{12} or C_9/C_9 products. These short-chain fatty acids subsequently are converted into their corresponding aldehydes and alcohols by the action of non-specific alcohol dehydrogenases. Ester formation from these alcohols is catalysed by acyl transferases, which are induced in ripening fruit (Perez *et al.*, 1996). In this way, many different odour derivatives are produced from relatively few precursors. We hypothesize that ACP levels increase in order to sustain the production of these volatile odour compounds that are lost from the fruit.

Sucrose transporter

Sugars make an important contribution to fruit flavour because they affect the balance between sweetness and acidity. The principal sugars in strawberry, fructose, glucose and sucrose show a steady increase as the fruit matures and ripens, reaching a total concentration of about 3% of the fresh weight. One of the ripening-enhanced clones we have isolated from strawberry encodes a putative sucrose transporter, a gene that has not been characterized previously in fruits. Sequence analysis indicates that the strawberry clone is substantially different from other sucrose transporters previously reported in source tissues such as leaves. Sucrose is the main form of carbohydrate translocated in the strawberry, and the sucrose transporter may have a key role in regulating the influx of this sugar into the fruit, a major sink tissue. We are investigating the biochemical properties of the recombinant protein expressed in yeast to identify the mechanism of sucrose transport through the membrane.

Invertase

The composition of the sugars in strawberry fruit will depend upon the rate of sucrose import, as determined by sink strength, and upon the rate of sucrose metabolism. Invertases have been studied in a number of fruits in relation to these processes, but little is known about the biochemical properties of the strawberry enzymes. Using degenerate primers, we have isolated two putative cell wall invertase clones from our ripe cDNA library and have investigated their expression in strawberry tissues by Northern analysis. Invertase transcripts were present at a low level in young expanding fruit, and decreased steadily as the fruit enlarged, falling to their lowest level in white fruit. Invertase expression increased again as the fruit began to ripen, with transcript levels attaining their highest values in over-ripe fruit. Cell wall forms of invertase are active in rapidly growing tissues and may facilitate growth by providing hexoses as a source of energy and carbon. The increase in invertase expression as the fruit ripens occurs as fructose, glucose and sucrose accumulate, indicating that this enzyme may be important in regulating the sweetness of the fruit. Confirmation of the function of these cell wall invertases in the strawberry awaits the analysis of transgenic fruit with altered levels of invertase expression.

Ripening Genes Affecting General Metabolism

Some of the ripening-enhanced genes identified in strawberry are likely to be involved in general metabolic processes in the fruit. These are described briefly below.

Cysteine (or thiol) proteinase

Cysteine proteinases occur widely in plants and are induced by various stress conditions, including cold, heat, salt and drought, and by wounding. The expression of cysteine proteinase genes is associated with ripening in a number of fruits including tomato and citrus, indicating that this enzyme may have a common role in fruit development. Senescing leaves also have elevated expression of this gene. The precise function of cysteine proteinases in fruits is uncertain, although they might be predicted to have a role in protein turnover, either by regulating the general rate of protein turnover or by activating specific proteins.

Elongation factor 2

This is another ripening-enhanced gene not previously reported in ripening fruits. Elongation factor 2 (EF-2) is an essential component of the translation machinery of eukaryotic cells, the expression of its gene correlating with rates of protein synthesis. It is a developmentally regulated gene in some systems, and may be important in controlling metabolic activity by regulating protein synthesis. Modification of the expression of this gene in fruit is expected to affect the rate of ripening.

Pyruvate decarboxylase

Little is known about the expression of pyruvate decarboxylase (PDC) in ripening fruits. The activity of PDC has been examined in tomato fruit in relation to ripening and anaerobic metabolism (Chen and Chase, 1993). There was no correlation with ripening, although the activity did show an increase in the fruit under hypoxic conditions. PDC is one of the more highly expressed of the strawberry ripening clones, its transcript being undetectable in mature green fruit but induced at 19 days post-anthesis, 3 days before the fruit turned colour.

PDC acts at an important step in glycolysis between aerobic and anaerobic metabolism and may divert pyruvate away from the respiratory pathway to form acetaldehyde and, subsequently, ethanol, via alcohol dehydrogenase. Therefore PDC could either regulate respiration, by affecting the availability of pyruvate to the mitochondria, or it could stimulate the biosynthesis of esters, which are important components of flavour.

Conclusions

The identity of several ripening-related genes from strawberry and their putative functions are described. We propose that they are involved in a broad spectrum of metabolic activity including cell wall hydrolysis, pigment and flavour biosynthesis, sugar accumulation and general functions related to protein turnover, respiration and the control of development. Although some of the strawberry genes identified here have been described in other fruits, several are unique to the strawberry and might be expected to affect fruit biochemistry in novel ways. We believe that the encoded biochemical functions of these genes are related to the quality characteristics of the fruit. The effects of modifying the expression of these genes on the biochemistry, composition and quality of strawberry fruit are being evaluated through the appropriate transgenic experiments.

Acknowledgements

This work was funded by the Biotechnology and Biological Sciences Research Council.

References

Chen, A.R.S. and Chase, T. (1993) Alcohol dehydrogenase 2 and pyruvate decarboxylase induction in ripening and hypoxic tomato fruit. *Plant Physiology and Biochemistry* 31, 855–875.

Colliver, S.P., Morris, P. and Robbins, M.P. (1997) Differential modification of flavonoid and isoflavonoid biosynthesis with an antisense chalcone synthase construct in transgenic *Lotus corniculatus*. *Plant Molecular Biology* 35, 509–522.

Given, N.K., Venis, M.A. and Grierson, D. (1988a) Purification and properties of phenylalanine ammonia-lyase activity from strawberry fruit and its synthesis during ripening. *Journal of Plant Physiology* 133, 25–30.

Given, N.K., Venis, M.A. and Grierson, D. (1988b) Hormonal regulation of ripening in the strawberry, a non-climacteric fruit. *Planta* 174, 402–406.

Given, N.K., Venis, M.A. and Grierson, D. (1988c) Phenylalanine ammonia-lyase activity and anthocyanin synthesis in ripening strawberry fruit. *Journal of Plant Physiology* 133, 25–30.

Harrison, R.E., Luby, J.J. and Furnier, G.R. (1997) Chloroplast restriction fragment variation among strawberry (*Fragaria* spp.) taxa. *Journal of the American Society for Horticultural Science* 122, 63-68.

Jain, S.M. and Pehu, E. (1992) The prospects of tissue culture and genetic engineering for strawberry improvement. *Acta Agriculturae Scandinavica, Section B, Soil and Plant Science* 42, 133-139.

Manning, K. (1991) Isolation of nucleic acids from plants by differential solvent precipitation. *Analytical Biochemistry* 195, 45-50.

Manning, K. (1993) Soft fruit. In: Seymour, G.B., Taylor, J.E. and Tucker, G.A. (eds), *Biochemistry of Fruit Ripening*. Chapman and Hall, London, pp. 347-377.

Manning, K. (1994) Changes in gene expression during strawberry fruit ripening and their regulation by auxin. *Planta* 194, 62-68.

Nitsch, J.P. (1950) Growth and morphogenesis of the strawberry as related to auxin. *American Journal of Botany* 37, 211-215.

Perez, A.G., Sanz, C., Olias, R., Rios, J.J. and Olias, J.M. (1996) Evolution of strawberry alcohol acyltransferase activity during fruit development and storage. *Journal of Agricultural and Food Chemistry* 44, 3286-3290.

Zabetakis, I. and Holden, M.A. (1997) Strawberry flavour: analysis and biosynthesis. *Journal of the Science of Food and Agriculture* 74, 421-434.

6 The Tomato Ethylene Receptor Gene Family: It's Not Easy Being a Plant

D. TIEMAN AND H. KLEE

Department of Horticultural Sciences, University of Florida, 1143 Fifield Hall, Gainesville, FL 32611, USA

Introduction

Ethylene, a gaseous plant hormone, plays an important regulatory role in many diverse plant developmental processes such as seed germination, fruit ripening, abscission and senescence (Abeles *et al.*, 1992). Ethylene is induced in response to most environmental stresses and usually causes a slowing of the plant's growth rate in less than optimal growth conditions. Ethylene is the plant hormone associated with regulating coordinated expression of many genes during ripening in climacteric fruits such as tomato. It is also essential for developmental processes such as abscission and flower petal senescence. While much of the regulation involving ethylene occurs at the level of synthesis, recent experiments characterizing the ethylene receptor have indicated that the signal transduction pathway for this hormone is also regulated. Researchers have speculated that hormone action must be mediated at the level of sensitivity (Trewavis, 1983; Bradford and Trewavis, 1994). Differential sensitivity can be exhibited in two contexts. First, adjacent cells in an organ can respond differentially to a hormone, as occurs during abscission. Second, the sensitivity of an organ to a hormone can change over time, as occurs during fruit ripening. How a plant regulates the variation in tissue or organ sensitivity to plant hormones spatially and temporally is not clear. Here, we describe our current knowledge of how sensitivity to ethylene changes through development of tomato fruit. Further, having access to genes that mediate ethylene sensitivity, we can ask whether we can take these genes and manipulate ethylene sensitivity in heterologous plant species, i.e. do components of the ethylene signal transduction pathway work in a 'mixed' system?

Ethylene Biosynthesis

The ethylene biosynthetic pathway is well characterized (McKeon and Yang, 1987). Ethylene is synthesized from methionine via S-adenosyl-L-methionine which subsequently is converted to 1-aminocyclopropane-1-carboxylate (ACC) by ACC synthase, the rate-controlling step in ethylene biosynthesis. This is followed by the conversion of ACC to ethylene by ACC oxidase. The ethylene biosynthetic pathway appears to be regulated at the level of ACC synthase gene transcription (Rottmann *et al.*, 1991; Liang *et al.*, 1992; O'Neill *et al.*, 1993), and ACC synthase genes are differentially regulated in response to various developmental and environmental stimuli (Rottmann *et al.*, 1991; Theologis, 1993; Kieber and Ecker, 1993). ACC oxidase activity, although present constitutively, is also regulated and may be responsible for the fine regulation of ethylene levels present in plant tissues (Kende, 1993).

Climacteric fruits, such as tomatoes, bananas and apples, undergo a burst of ethylene evolution and respiration at the onset of ripening, while non-climacteric fruits, such as citrus, do not. In climacteric fruits, ethylene promotes ripening by coordinately inducing the expression of many genes whose products are involved in the ripening process (Slater *et al.*, 1985; Lincoln and Fischer, 1988a, b). Ripening is associated with breakdown of chlorophyll, cell wall breakdown, resulting in fruit softening, synthesis of aromatic compounds, and conversion of starch to sugars. Inhibition of ethylene synthesis can disrupt these processes, resulting in fruit with delayed ripening characteristics. Ethylene synthesis can be reduced effectively by the expression of antisense genes for either ACC synthase (Oeller *et al.*, 1991) or ACC oxidase (Hamilton *et al.*, 1990) or by expression of a bacterial ACC-degrading enzyme, ACC deaminase (Klee *et al.*, 1991). In ethylene-inhibited fruit, the time that it takes to proceed from breaker stage (the stage at which the fruit first exhibits external colour change) to full red is inversely proportional to the rate of ethylene synthesis (H. Klee *et al.*, unpublished). The most extreme example is illustrated by the antisense ACC synthase fruit (Oeller *et al.*, 1991) that are 99% inhibited for ethylene synthesis and never fully ripen without exogenous ethylene application. Progressively higher rates of ethylene synthesis result in faster and more complete ripening. Although the introduction of any of these genes yields fruit that are delayed in ripening, the antisense ACC synthase and ACC oxidase genes are more likely to produce extreme fruit phenotypes that never fully ripen. Introduction of the ACC deaminase gene is less effective in the complete elimination of ethylene synthesis, presumably because the enzyme is competing with the fairly abundant ACC oxidase for substrate. However, lines with extremely reduced ethylene levels are not necessarily commercially desirable since they require continuous exposure to high levels of ethylene over an extended period to achieve full ripening. Thus, the actual commercial target is a fruit that is only partially reduced in ethylene synthesis and will exhibit delayed ripening without the necessity for prolonged applications of ethylene for ripening. With this conclusion in mind, it appears that any of these three transgenes should deliver a commercially useful extended shelf life fruit.

Although ethylene acts to coordinate and hasten the ripening process of climacteric fruits, several lines of evidence suggest that other levels of control precede the increase in ethylene synthesis. Experiments conducted on avocado (Eaks, 1980), apple (Harkett *et al.*, 1971) and tomato (Yang, 1987) indicate that immature fruits do not respond to ethylene by ripening. These fruits do perceive ethylene since some ethylene-inducible genes such as ACC oxidase are activated when fruits are exposed to ethylene, but they do not initiate the developmental sequence that leads to ripening. Even in mature tomato fruits that ripen more quickly when exposed to exogenous ethylene, ripening does not initiate uniformly. Rather, it proceeds from the locules to the pericarp and from the blossom end to the stem end. Since ethylene is readily diffusible within the fruit, a reasonable explanation for asynchronous ripening is differential regulation of ethylene sensitivity. Thus, it appears that not only do different plant tissues respond differentially to ethylene, but fruit and fruit tissues can respond to exogenous ethylene in different ways depending on their developmental state.

Differential developmental control of ethylene sensitivity is not limited to fruits. In many flowers, tissue responsiveness to ethylene increases with maturation, and the mechanisms controlling the differential responsiveness are likely to be similar, if not identical, to those controlling fruit ripening. In the tomato *Nr* mutant, flower senescence and abscission are affected along with fruit ripening, indicating that a single gene can control these processes. Typically, flowers begin to respond to ethylene at the time that the stigma becomes receptive to pollen. For example, young carnation flowers are essentially unresponsive to exogenous ethylene, while mature flowers exhibit petal inrolling and older flowers respond by wilting and initiating autocatalytic ethylene synthesis (Barden and Hanan, 1972). Similar development of ethylene responsiveness can be observed in geranium flowers (Evensen, 1991).

Ethylene Perception and Signal Transduction in *Arabidopsis*

The ease of growth and genetic manipulation of *Arabidopsis* has made it a popular research tool for understanding plant growth and developmental processes. *Arabidopsis* has also proven invaluable in understanding and identifying genes involved in plant responses to ethylene. Most of our knowledge of the ethylene signal transduction pathway is based on genetic analyses performed in *Arabidopsis*. The combination of excellent genetics and a simple and reliable assay has resulted in identification of a large number of ethylene-insensitive mutants. Several genes involved in the ethylene signal transduction pathway have been identified by screening dark-grown *Arabidopsis* seedlings for the lack of a normal 'triple response' when grown in the presence of ethylene. Etiolated seedlings grown in the presence of ethylene will respond by inhibition of root and hypocotyl elongation, radial swelling of the hypocotyl, and an exaggerated apical hook. Ethylene-insensitive mutants, in contrast, behave like seedlings germinated in the absence of ethylene or ACC, with narrow elongated hypocotyls with no apical hook and long roots. Several *Arabidopsis* mutants that do not show this response and appear as if grown in the absence of ethylene were identified. Several of the genes involved in the ethylene signal transduction pathway, including *ETR1*, have been isolated using positional cloning. Dominant mutations in the *ETR1* (ethylene responsive 1) gene and recessive mutations in *EIN2* (ethylene insensitive 2) and *EIN3* result in a lack of the triple response in response to exogenous ethylene. Mutations in the *CTR1* (constitutive response 1) gene result in a constitutive ethylene response phenotype even in the absence of ethylene. Double mutant analysis has shown that *CTR1* is epistatic to *ETR1* and *EIN4*, while *EIN2* and *EIN3* act after *CTR1* (Ecker, 1995; Roman *et al.*, 1995). Each of these genes has been identified in *Arabidopsis*, and molecular analysis of these genes should further our understanding of the mechanisms of ethylene signal transduction. CTR1 has been shown to have homology to the Raf family of serine/threonine protein kinases (Kieber *et al.*, 1993). The EIN3 protein has been localized to the nucleus, and appears to be a member of a gene family in *Arabidopsis* (Chao *et al.*, 1997). The ETR1 protein binds ethylene when expressed in yeast, indicating that it is an ethylene receptor (Schaller and Bleecker, 1995). How these proteins interact to transduce the ethylene signal and ultimately result in a cellular response should become clearer as more genes in the pathway are identified.

ETR1 and its Homologues in *Arabidopsis*

Recent developments have focused on what appears to be the first component of the ethylene signal transduction cascade, *ETR1*. All of the mutant alleles of *ETR1* are dominant and genetic analyses indicate that it is at or near the earliest step in ethylene perception. The *etr-1* mutant is defective in all ethylene responses that have been examined. The gene encoding *ETR1* was isolated using map-based cloning (Chang *et al.*, 1993). Sequence comparisons indicate that the C-terminal end of ETR1 is similar to a class of prokaryotic proteins involved in signal transduction referred to as two-component histidine kinases, while the N terminus of ETR1 is unique among proteins in the databases. In bacteria, this class of proteins is responsible for environmental sensing for the presence of nutrients or changes in osmolarity (Parkinson, 1993).

By analogy to the bacterial two-component proteins, the ETR1 protein can be divided into three distinct domains, represented schematically in Fig. 6.1. The first domain, consisting of approximately amino acids 1–313, encodes a sensor domain that is responsible for ligand binding. This domain contains three hydrophobic, potential membrane-spanning regions. All of the known mutations in *ETR1* causing ethylene insensitivity lie within this region of the protein. The second ETR1 domain encodes a putative autophosphorylating histidine kinase. Finally, the third domain, called either the receiver or response regulator, contains an aspartic acid residue that can act as a receiver for the phosphate from the histidine kinase domain. Although the third domain can act as a phosphate receiver, in bacteria it is not responsible for transducing the signal from the histidine kinase to the next step in the phosphorylation cascade. Indeed, the response regulator domain is not even present in some two-component systems (Winans *et al.*, 1994). Rather, it may act as a modulator of phosphate transfer to the actual receiver, which is on a separate protein. In this

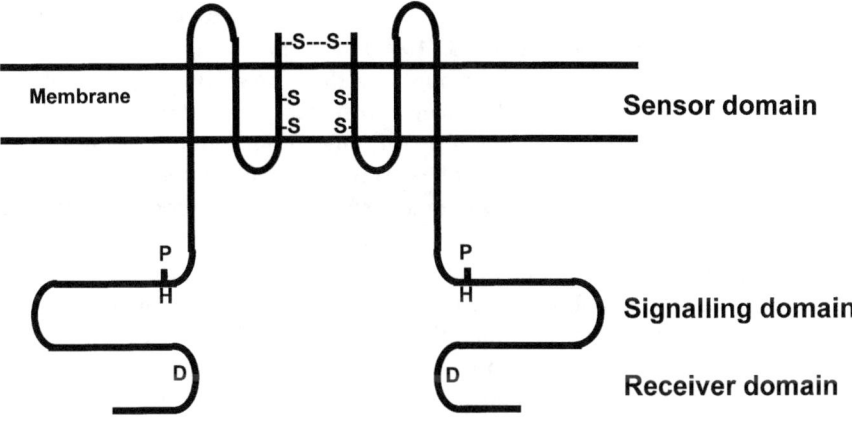

	Full protein	Sensor membrane spanning	Signalling histidine kinase	Receiver response regulator
At-ERS	79/68	82/72	77/64	
At-ETR2	64/40	71/49	57/32	65/41
At-EIN4	61/39	66/48	52/29	69/39
Le-NR	81/68	85/73	81/67	
Le-ETR1	90/82	96/92	88/78	85/70
Le-ETR2	89/78	93/87	87/76	81/59
Le-ETR4	66/42	70/51	62/34	65/37
Le-ETR5	64/40	67/50	56/32	67/38

Fig. 6.1. Schematic diagram of the domains of the *Arabidopsis* ETR1 protein and amino acid identities and similarities among the five known tomato and four *Arabidopsis* ETR1 homologous proteins. The top panel is a representation of the *Arabidopsis* ETR1 protein with the three domains indicated. The sensor domain contains three regions capable of spanning a membrane. The signalling domain contains the histidine kinase region that becomes phosphorylated and then transfers the phosphate to the aspartate of an unknown substrate. The receiver domain contains an aspartate (D) that is capable of receiving the phosphate from the histidine kinase. The bottom panel shows the amino acid identity and similarity, respectively, between the *Arabidopsis* ETR1 protein and each of the published tomato (Le) and *Arabidopsis* (At) ETR1 homologues within each of the three domains and along the full protein sequence.

way, the response regulator domain might act as a buffer for signal transduction by acting as a dead end for phosphate transfer. An *Arabidopsis* gene lacking the response regulator domain, designated ERS, has since been cloned. Introduction of a mutation in the membrane-spanning domain of ERS confers ethylene insensitivity in transgenic *Arabidopsis*. Two other *Arabidopsis* ETR1 homologues, *EIN4* and *ETR2*, have also been identified. Mutations in the membrane-spanning domain of these two proteins result in ethylene insensitivity, as well (Hua *et al.*, 1997). Each of these proteins contains the third response regulator domain that is missing in the ERS protein.

In bacteria, the histidine kinases act as dimers, with one subunit phosphorylating the other. Schaller *et al.* (1995) have shown that ETR1 also forms dimers. This dimerization could account for the observation that all of the known mutations in ETR1 are dominant. Presumably, a mutant subunit of ETR1 can inactivate a wild-type subunit by forming an unproductive dimer. While such a model can explain the data, it has yet to be demonstrated conclusively. Isolation of only dominant mutations could also be due to functional redundancy in the receptor gene family (see below). It is also possible that the ETR1 protein acts to negatively regulate the pathway in the absence of ethylene. In this model, ethylene would bind to ETR1, leading to shut-off of the protein and a derepression of the pathway. A mutant unable to bind ethylene would be locked in the 'on' position, effectively preventing derepression. Definitive biochemical proof that the *ETR1* gene encodes an ethylene receptor has been obtained. Schaller and Bleecker (1995) expressed the ETR1 protein in yeast cells and demonstrated that the protein binds ethylene. The dosage-dependent binding in yeast parallels the curve for growth inhibition responses for ethylene in *Arabidopsis*. Ethylene binding was localized to the N-terminal hydrophobic domain and was eliminated in the ethylene-unresponsive *etr1-1* mutant protein.

The Tomato *ETR1* Gene Family

Never-ripe

While a great deal of effort has gone into understanding ethylene synthesis during fruit ripening, very little attention has been paid to ethylene perception. We initiated a search for tomato mutants that were altered in their ability to perceive ethylene. *Never-ripe* (*Nr*) is a semi-dominant mutation originally identified by the inability of its fruit to undergo ripening. Our analysis of *Nr* indicated a number of pleiotropic effects indicative of ethylene insensitivity throughout the plant (Lanahan *et al.*, 1994). Dark-grown seedlings do not show the classic triple response when grown in the presence of ethylene. The mutant is greatly impaired in pedicel abscission. There are also significant delays in leaf and flower petal senescence. The *Nr* fruit never fully ripen and are phenotypically indistinguishable from the best ACC synthase antisense line, A11.1, described by Oeller *et al.* (1991).

Efforts were next focused on identification of the *ETR1* homologous sequences in tomato with an emphasis on the *Nr* mutant. The *Arabidopsis ETR1* genomic clone was used to probe Southern blots of DNA from an F_2 segregating population derived from a cross between *Lycopersicon esculentum* and *L. pinnellii*. Five independent hybridizing sequences mapping to chromosomes 7, 9, 10, 11 and 12 were identified (Yen *et al.*, 1995). It has yet to be demonstrated that all five loci contain functional *ETR1* homologous sequences. However, the first three genes that have been cloned all contain very significant *ETR1* sequence identity. What was particularly significant about the mapping is that one of the loci is tightly linked to the *Nr* mutation on chromosome 9, within 0.8

cM. The phenotypic similarity of *etr1-1* and *Nr* strongly suggested that the molecular basis of *Nr* is a mutation in an *ETR1* homologue.

Screening of a ripening tomato fruit cDNA library was accomplished by hybridization with the full-length *Arabidopsis ETR1* genomic clone. A full-length cDNA encoding a protein with similarity to ETR1 was identified. A gene-specific hybridization probe enabled determination of the map location as being chromosome 9, near the *Nr* locus. The predicted protein encoded by this cDNA is 68% identical and 81% similar to the *ETR1* sequence. However, the protein lacks the response regulator domain found in ETR1 (Fig. 6.1). Significantly, all four of the amino acids that, when changed, result in dominant mutations in the *Arabidopsis* gene are conserved in the tomato protein. DNA sequence comparisons of the tomato *Nr* mutant and the isogenic wild-type gene revealed a single nucleotide change that causes a proline to leucine switch at amino acid 36 in the Nr mutant protein. This proline is conserved in the *Arabidopsis* ETR1, EIN4, ERS and ETR2 proteins. When the mutant tomato cDNA under the control of the constitutive cauliflower mosaic virus (CaMV) 35S promoter was introduced into wild-type tomato plants, the resulting transgenic plants exhibited ethylene insensitivity, conclusively demonstrating that the *Nr* phenotype is caused by a mutation in this ETR1-related member of the tomato ethylene receptor family (Wilkinson *et al.*, 1995).

The NR protein of tomato resembles the *Arabidopsis* ERS in that it lacks the response regulator domain found in the ETR1 protein. Conversely, we have characterized four additional tomato gene family members, Le-ETR1, Le-ETR2, Le-ETR4 and Le-ETR5, all of which contain response regulator domains (Fig. 6.1). Thus, the family of ethylene receptors appears to be complex both in structure and number.

We used the *NR* cDNA as a probe to examine the expression pattern of *NR* mRNAs during fruit ripening. Ethylene inducibility in mature green fruit was also examined. We found that the *NR* mRNA is both developmentally regulated and ethylene inducible. *NR* gene expression increases substantially from green to red ripe stages of fruit ripening and is reduced substantially in the *Nr* mutant. *NR* mRNA expression is induced by ethylene treatment in mature green fruits but not immature green fruits. The data are consistent with a model in which *NR* gene expression is part of the feedback induction that occurs during ripening. Thus, not only would ethylene synthesis be positively feedback regulated, but ethylene perception would also be up-regulated in response to ethylene. At this time, it cannot be determined whether higher levels of receptor would have a positive or a negative effect on ethylene response since the NR protein could either activate or suppress downstream signal transduction components but ripening occurs as levels of the receptor increase.

The existence of a class of genes, represented by *NR* and E8, that are ethylene regulated in a developmentally dependent manner illustrates the fundamental change in ethylene sensitivity that must occur upon achieving competency to ripen. That the ethylene receptor itself is regulated in this manner is particularly intriguing. Developmental regulation of the ethylene signal transduction pathway would provide a molecular basis to explain the early observations of differential ethylene responsiveness of immature vs.

mature tomato fruits. It is important to note that expression of *NR* is not precisely like that of E8. While ethylene inducibility is developmentally controlled, there is a detectable basal level of expression of *NR* in other tissues. However, when the fruit must express many genes rapidly and coordinately, the *NR* gene product is greatly increased. One possible explanation for the differential ethylene responsiveness of immature and mature fruit tissue may be quantitative. The components of ethylene signal transduction may be rate limiting to some, but not all, ethylene-regulated genes in immature fruits. Also, different ETR1 homologous proteins may affect different pathways. While this concept must still be tested, it does appear that we must expand the concept of climacteric autocatalytic ethylene synthesis to include the ethylene receptor.

Other tomato *ETR1* homologues

The importance of ethylene perception in plant responses to the environment and fruit ripening has led us in the search for tomato ethylene receptor homologues. Mapping of *ETR1* homologues on the tomato genome has shown that as many as five hybridizing loci can be found (Yen *et al.*, 1995). We and others have cloned two genes, *Le-ETR1* and *Le-ETR2*, homologous to the *ETR1* gene of *Arabidopsis* and the *NR* gene of tomato (Zhou *et al.*, 1996a, b). *Arabidopsis* has been shown to have at least four *ETR1*-like genes in which mutations in the membrane-spanning domains can result in ethylene insensitivity. In order to identify more genes that may be involved in ethylene perception in plants, we have also used the *Arabidopsis ETR2* cDNA (Hua *et al.*, 1997) as a probe to identify two additional tomato genes, designated *Le-ETR4* and *Le-ETR5*. *Le-ETR4* was 59% similar to *ETR2*, while *Le-ETR5* was 62% similar to *ETR2* at the nucleotide level. Le-ETR4 was 78% similar and 60% identical to the ETR2 protein at the translated amino acid level, while Le-ETR5 was 76% similar and 54% identical at the amino acid level. Each of these proteins contains the response regulator domain absent from the *NR* protein. The *Arabidopsis* ETR1 protein has been divided into three domains; the membrane-spanning (sensor) domain the histidine kinase (signalling) domain and the response regulator (receiver) domain. Figure 6.1 shows the amino acid identities and similarities of each of the *Arabidopsis* and tomato ETR1 homologues compared with the *Arabidopsis* ETR1 ethylene receptor. The similarities among these genes is most striking in the membrane-spanning domain, while the histidine kinase and response regulator domains are more divergent. Tomato Le-ETR1 and Le-ETR2 are the most similar to ETR1, while tomato Le-ETR4 and Le-ETR5 and *Arabidopsis* ETR2 are the least similar. Only the tomato NR and *Arabidopsis* ERS proteins are missing the response regulator domain. Interestingly, the putative autophosphorylated histidine in the histidine kinase domain is present in all the published tomato and *Arabidopsis* homologues except ETR2 and Le-ETR5. The aspartate which has been suggested to act as a phosphate receiver from the autophosphorylated histidine is present in all the proteins, except ERS and NR, which do not have the response regulator domain. Also of interest is the conservation in all tomato and *Arabidopsis* proteins of amino acids which, if altered, are responsible for the formation of the known dominant ethylene-insensitive mutants of *Arabidopsis*

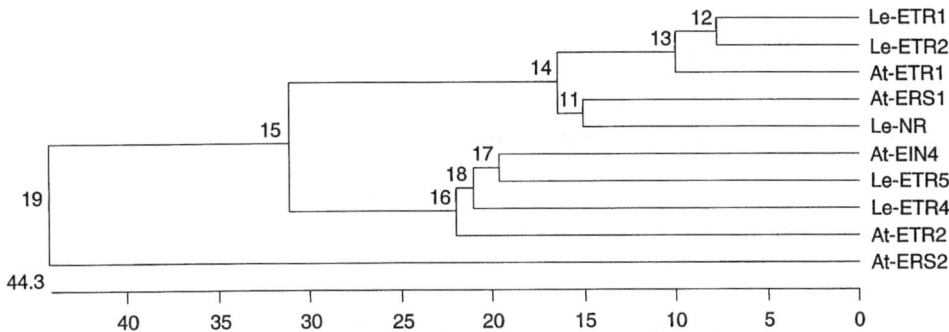

Fig. 6.2. Phylogenetic tree analysis of the five tomato (Le) and five *Arabidopsis* (At) ETR1 homologous proteins. Protein sequences were analysed using the CLUSTAL method.

or tomato. Other invariant amino acids include the two cysteines involved in covalent dimerization (at positions 4 and 6 in ETR1) of the ETR1 protein (Schaller *et al.*, 1995), although the intervening amino acid is missing in Le-ETR5. The membrane-spanning regions thought to be important in ethylene binding are well conserved in all the tomato and *Arabidopsis* proteins.

Phylogenetic tree analysis of the *Arabidopsis* and tomato deduced amino acid sequences of the ETR1 homologues indicates that Le-ETR1 and Le-ETR2 are most closely related to ETR1 of *Arabidopsis*, while NR is most closely related to the ERS protein of *Arabidopsis* (Fig. 6.2). Le-ETR4 and Le-ETR5 are more closely related to *Arabidopsis* ETR2 and EIN4. The known tomato and *Arabidopsis* genes can be divided into three categories using phylogenetic analysis and what is known of their function in transgenic plants. The first category, represented by *Le-ETR1*, *Le-ETR2* and *ETR1*, contains genes that have the third response regulator domain. The second group consists of genes that do not contain the response regulator domain (*NR* and *ERS1*). The third group consists of *ETR2*, *EIN4*, *Le-ETR4* and *Le-ETR5*. *Le-ETR4* and *Le-ETR5* had not been identified earlier in screenings of tomato cDNA or genomic libraries with *ERS*, *ETR1*, *NR*, *Le-ETR1* or *Le-ETR2* as probes at low stringency, indicating their relative divergence from the category 1 and 2 genes. Phylogenetic analysis also indicated that these genes are divergent from the other tomato and *Arabidopsis* genes (Fig. 6.2).

It is interesting to note that of all the known tomato and *Arabidopsis* ethylene receptor-like proteins only NR, ERS2 and ERS do not have the third response regulator domain. The importance of this domain in transmitting the ethylene signal down the signal transduction pathway is unknown. In bacterial two-component sensing systems, the response regulator domain, which is often on a separate protein, receives the phosphate from the histidine kinase domain, from which the phosphate is transmitted down the signal transduction cascade. Whether a separate response regulator protein is present in plants is unknown. Alternately, the ethylene signal could be transmitted from NR or ERS to the response regulator portion of an ETR1 homologous protein containing the response regulator domain. Since the proteins are known to form dimers in the membrane, heterodimers between proteins with and without the response regulator domain are possible.

Transgenic Plants Altered in Ethylene Perception

We have introduced both the mutant *Nr* gene and the *Arabidopsis etr1-1* mutant cDNA under the control of the CaMV 35S promoter into tomato plants and assessed the effects of the transgenes on the ability of these plants to respond to ethylene. Our logic in undertaking these parallel experiments was that the heterologous *Arabidopsis* protein might not interact effectively with the next downstream component of ethylene signal transduction and that the homologous tomato gene might be necessary for effective control. This is not the case. Both the *Arabidopsis etr1-1* (Wilkinson *et al.*, 1997) and the tomato *Nr* mutant (Wilkinson *et al.*, 1995) genes work effectively in transgenic tomatoes to confer ethylene insensitivity. Transgenic plants expressing the *Arabidopsis etr1-1* mutant cDNA are phenotypically indistinguishable from the *Nr* mutant plants in terms of seedling triple response, flower abscission and petal wilting, and fruit ripening.

As further proof that the *Arabidopsis* mutant gene could work in a heterologous plant species, the gene has been introduced into petunia plants (Wilkinson *et al.*, 1997). In petunia, flower corolla wilting typically occurs within 48 h after pollination and is known to be an ethylene-mediated response to a pollination-associated signal. Transgenic plants expressing the *etr1-1* gene exhibited no ethylene-induced corolla wilting for at least 8 days either following pollination or in response to exogenously applied ethylene. Thus, it appears that the transgene will be effective in increasing shelf life substantially for many floricultural crops that senesce in response to ethylene.

Conclusions

There is clearly hormonal regulation of fruit development at the level of perception and strong evidence that regulation of ethylene perception is a major control point in the commitment to fruit ripening. The tomato fruit is an excellent model for differential regulation of hormone perception during development. It has long been postulated that phytohormone effects could be modulated at the level of tissue sensitivity. We can state that the concept of autocatalytic ethylene regulation must be modified to include not only synthesis of the hormone but also modulation of the receptor. As our understanding of the regulation of the tomato ethylene receptor family advances, it should be possible to define unambiguously the role of ethylene and ethylene perception in fruit ripening and senescence. We have identified five tomato genes with homology to the *Arabidopsis ETR1* gene. We are now conducting experiments to determine the role of the additional tomato *ETR1* gene family members in the diverse plant responses to ethylene. Understanding the regulation of the ethylene signal transduction pathway will lead to an ability to block processes in plant development selectively to create plants with greater economic value.

References

Abeles, F.B., Morgan, P.W. and Saltveit, M.E. (1992) *Ethylene in Plant Biology*. Academic Press, San Diego.

Barden, L.E. and Hanan, J.J. (1972) Effect of ethylene on carnation keeping life. *Journal of the American Society of Horticultural Science* 97, 785-788.

Bradford, K. and Trewavis, A. (1994) Sensitivity thresholds and variable time scales in plant hormone action. *Plant Physiolology* 105, 1029-1036.

Chang, C., Kwok, S.F., Bleecker, A.B. and Meyerowitz, E.M. (1993) *Arabidopsis* ethylene-response gene *ETR1*: similarity of products to two-component regulators. *Science* 262, 539-544.

Chao, Q., Rothenberg, M., Solano, R., Roman, G., Terzaghi, W. and Ecker, J.R. (1997) Activation of the ethylene gas response pathway in *Arabidopsis* by the nuclear protein ETHYLENE-INSENSITIVE3 and related proteins. *Cell* 89, 1133-1144.

Eaks, I.L. (1980) Respiratory rate, ethylene production and ripening response of avocado fruit to ethylene or propylene following harvest at different maturities. *Journal of the American Society of Horticultural Science* 105, 744-747.

Ecker, J.R. (1995) The ethylene signal transduction pathway in plants. *Science* 268, 667-675.

Evensen, K.B. (1991) Ethylene responsiveness changes in *Pelargonium* × *domesticum* florets. *Physiologia Plantarum* 82, 409-412.

Hamilton, A.J., Lycett, G.W. and Grierson, D. (1990) Antisense gene that inhibits synthesis of the hormone ethylene in transgenic plants. *Nature* 346, 284-287.

Harkett, P., Hulme, A., Rhodes, M. and Wooltorton, L. (1971) The threshold value for physiological action of ethylene on apple fruits. *Journal of Food Technology* 6, 39-45.

Hua, J., Sakai, H. and Meyerowitz, E.M. (1997) The ethylene receptor gene family in *Arabidopsis*. In: Kanellis, A.K., Chang C., Kende H. and Grierson D. (eds) *Biology and Biotechnology of the Plant Hormone Ethylene*. Kluwer Academic Publishers, Dordrecht, pp. 71-76.

Kende, H. (1993) Ethylene biosynthesis. *Annual Review of Plant Physiology and Plant Molecular Biology* 44, 283-307.

Kieber, J.J. and Ecker, J.R. (1993) Ethylene gas, it's not just for ripening anymore. *Trends in Genetics* 9, 356-362.

Kieber, J.J., Rothenberg, M., Roman, G., Feldmann, K.A. and Ecker, J.R. (1993) *CTR1*, a negative regulator of the ethylene response pathway in *Arabidopsis*, encodes a member of the raf family of protein kinases. *Cell* 72, 427-441.

Klee, H.J., Hayford, M.B., Kretzmer, K.A., Barry, G.F. and Kishore, G.M. (1991) Control of ethylene synthesis by expression of a bacterial enzyme in transgenic tomato plants. *The Plant Cell* 3, 1187-1193.

Lanahan, M.B., Yen, H.-C., Giovannoni, J.J. and Klee, H.J. (1994) The *Never ripe* mutation blocks ethylene perception in tomato. *The Plant Cell* 6, 521-530.

Liang, X., Abel, S., Keller, J.A., Shen, N.F. and Theologis, A. (1992) The 1-aminocyclopropane-1-carboxylate synthase gene family of *Arabidopsis thaliana*. *Proceedings of the National Academy of Sciences USA* 89, 11046-11050.

Lincoln, J.E. and Fischer, R.L. (1988a) Regulation of gene expression by ethylene in wild-type and *rin* tomato (*Lycopersicon esculentum*) fruit. *Plant Physiology* 88, 370-374.

Lincoln, J.E. and Fischer, R.L. (1988b) Diverse mechanisms for the regulation of ethylene-inducible gene expression. *Molecular and General Genetics* 212, 71-75.

McKeon, T. and Yang, S.F. (1987) Biosynthesis and metabolism of ethylene. In: Davies P.

(ed.), *Plant Hormones and Their Role in Plant Growth and Development*. Martinus Nijhoff, Boston, Massachusetts, pp. 94-112.

Oeller, P.W., Min-Wong, L., Taylor, L.P., Pike, D.A. and Theologis, A. (1991) Reversible inhibition of tomato fruit senescence by antisense RNA. *Science* 254, 437-439.

O'Neill, S., Nadeau, J.A., Zhang, X.S., Bui, A.Q. and Halevy A.H. (1993) Interorgan regulation of ethylene biosynthetic genes by pollination. *The Plant Cell* 5, 419-432.

Parkinson, J. (1993) Signal transduction schemes of bacteria. *Cell* 73, 857-871.

Roman, G., Lubarsky, B., Kieber, J., Rothenberg, M. and Ecker, J. (1995) Genetic analysis of ethylene signal transduction in *Arabidopsis thaliana*: five novel mutant loci integrated into a stress response pathway. *Genetics* 139, 1393-1409.

Rottmann, W.H., Peter, G.F., Oeller, P.W., Keller, J.A., Shen, N.F., Taylor, L.D., Campbell, A.D. and Theologis, A. (1991) 1-Aminocyclopropane-1 carboxylate synthase in tomato is encoded by a multigene family whose transcription is induced during fruit and floral senescence. *Journal of Molecular Biology* 222, 937-961.

Schaller, G.E. and Bleecker, A.B. (1995) Ethylene binding sites generated in yeast expressing the *Arabidopsis ETR1* gene. *Science* 270, 1809-1811.

Schaller, G.E., Ladd, A.N., Lanahan, M.B., Spanbauer, J.M. and Bleecker, A.B. (1995) The ethylene response mediator ETR1 from *Arabidopsis* forms a disulfide-linked dimer. *Journal of Biological Chemistry* 270, 12526-12530.

Slater, A., Maunders, M.J., Edwards, K., Schuch, W. and Grierson D. (1985) Isolation and characterization of cDNA clones for tomato polygalacturonase and other ripening-related proteins. *Plant Molecular Biology* 5, 137-147.

Theologis A. (1993) One rotten apple spoils the whole bushel: the role of ethylene in fruit ripening. *Cell* 70, 181-184.

Trewavis A. (1983) Is plant development regulated by changes in the concentration of growth substances or by changes in the sensitivity to growth substances? *Trends in Biochemical Sciences* 8, 354-357.

Wilkinson, J.Q., Lanahan, M.B., Clark, D.G., Bleecker, A.B, Chang, C., Meyerowitz, E.M. and Klee, H.J. (1997) A dominant mutant receptor from *Arabidopsis* confers ethylene insensitivity in heterologous plants. *Nature Biotechnology* 15, 444-447.

Wilkinson, J.Q., Lanahan, M.B., Yen, H.-C., Giovannoni, J.J. and Klee H.J. (1995) An ethylene-inducible component of signal transduction encoded by *Never-ripe*. *Science* 270, 1807-1809.

Winans, S., Mantis, J., Chen, C., Chang, C. and Han, D. (1994) Host recognition by the virA, virG two-component regulatory proteins of *Agrobacterium tumefaciens*. *Research in Microbiology* 145, 461-473.

Yang, S.F. (1987) The role of ethylene and ethylene synthesis in fruit ripening. In: Thompson, W., Nothnagel, E. and Huffaker, R. (eds), *Plant Senescence: Its Biochemistry and Physiology*. The American Society of Plant Physiologists, Rockville, Maryland, pp. 156-165.

Yen, H.-C., Lee, S., Tanksley, S., Lanahan, M., Klee, H., and Giovannoni, J. (1995) The tomato *Never-ripe* locus regulates ethylene-inducible gene expression and is linked to a homologue of the *Arabidopsis ETR1* gene. *Plant Physiology* 107, 1343-1353.

Zhou, D., Kalaitzis, P., Mattoo, A.K. and Tucker, M.L. (1996a) The mRNA for an ETR1 homologue in tomato is constitutively expressed in vegetative and reproductive tissues. *Plant Molecular Biology* 30, 1331-1338.

Zhou, D., Mattoo, A.K. and Tucker, M.L. (1996b) Molecular cloning of a tomato cDNA encoding an ethylene receptor. *Plant Physiology* 110, 1435-1436.

7 Environmental Requirements as Determined by Rooting Potential in Leafy Cuttings

R.S. Harrison-Murray and B.H. Howard

Horticulture Research International, Maidstone, Kent ME19 6BJ, UK

Introduction

Importance of vegetative propagation

Many horticultural crops are selected for particular characteristics which must be reproduced faithfully by vegetative propagation, in which particular organs are induced to develop into complete plants. Stem cuttings are used most commonly, with the need to induce adventitious roots. Leafy 'softwood' cuttings propagated mainly in early summer are the most popular type, with estimates of up to 200 million cuttings of mainly woody ornamental species being propagated each year by the UK hardy nursery stock sector alone; 70% of the propagation systems used in the USA depend upon the successful rooting of cuttings (Davies *et al.*, 1994).

Propagation by leafy stem cuttings is a key stage in the development of new varieties of woody and herbaceous species, with the need for success greater than ever before as new but costly methods of plant improvement supplement conventional breeding. In this context, it is important to realize that in seeking improvements to flower, fruit and foliage characteristics, it is often the case that little or no attention is given to the rooting ability of the cuttings, and often potentially valuable new varieties are underexploited commercially because they cannot be propagated conveniently or cost-effectively. Propagation then becomes a technical challenge, and this in turn emphasizes the need for continually improved understanding of key processes in the initiation and development of adventitious roots, leading to the development of even more effective propagation techniques compared with those available today.

Nature of the crop

A 'crop' of softwood cuttings is unlike any other horticultural crop in that it comprises parts of leafy stems, which are often the most immature distal parts used to provide 'tip' or 'apical' cuttings. When collected from the stockplant, these cuttings are removed from their natural water supply and are therefore acutely vulnerable to water stress. Water deficits may develop during preparation of the cuttings, which involves trimming the stem, removal of some of the lower leaves and application of other treatments (Macdonald, 1986). More often it occurs after the cuttings have been inserted into the rooting medium. The ability of cuttings to take up and conduct water from the rooting medium is limited, partly by resistance to water movement through the medium and at the interface with the cutting (Thomas and Harrison-Murray, 1995), and partly because the conductance of the vascular tissue declines rapidly, presumably due to the vascular elements becoming occluded by air, detritus, microbial material and possibly tyloses (Grange and Loach, 1983). As a result, survival depends on a restriction of transpiration to a similar degree, which for most cuttings can only be achieved by provision of an aerial environment in which evaporative demand is low.

During the period of root initiation and development, cuttings must receive sufficient light to avoid serious depletion of their carbohydrate reserves, and preferably they should photosynthesize at a rate which exceeds respirational depletion, as shown below. However, causal relationships between carbohydrate balance and adventitious rooting remain obscure (Veierskov, 1988), and net changes in carbohydrate, conveniently measured as dry weight changes in unrooted cuttings, show general rather than precise correlation with rooting. Root production may be proportional to dry matter increases rather than monopolizing total carbohydrate gains, and is not necessarily prevented by a net loss of dry matter, suggesting that root formation is able to compete successfully with alternative sinks for carbohydrate (Howard, 1965).

Only a small proportion of the radiant energy absorbed by leaves is used in photosynthesis, the remainder being converted into heat which increases leaf temperature and thereby stimulates transpiration. Therefore, until roots form, the key requirement is to ensure that cuttings receive sufficient photosynthetically active radiation (PAR) to maximize carbon balance whilst avoiding levels of irradiance that would lead to severe water stress (Davis, 1988).

The 'tools' available to the practical propagator are described elsewhere (e.g. Hartmann *et al.*, 1990) and include synthetic auxins such as indolebutyric acid (IBA) and the ability to heat the rooting medium (bottom heat), both of which are aimed at stimulating the process of root initiation and increasing the speed of root development. Shade is used to limit the amount of radiant energy reaching the cutting, but is rarely adjusted so that irradiance varies greatly in response to diurnal and weather variations. Polyethylene film, either laid directly on the cuttings, or used to form a 'tent' over a bench or in the form of larger 'polyhouses', is used to raise humidity. Further humidification can be achieved by injecting fine droplets of water (fog) a proportion of which are small enough to remain suspended in air indefinitely. More common is the use of relatively coarse droplets from 'mist' nozzles, which settle rapidly and serve mainly to keep the foliage wet rather than to humidify directly. Any distinction between mist and fog on the basis of droplet size is rather arbitrary, especially since fine droplets tend to coalesce when at high concentration, creating larger droplets which settle out on to the foliage and other surfaces. The term 'wet fog' implies a system with a range of droplet sizes which ensures some wetting of the leaves combined with effective humidification.

Water balance of cuttings

Avoidance of damaging water deficits is a basic requirement for successful rooting of cuttings though the precautions needed to achieve this vary between species. Here, we review briefly our understanding of the principles involved.

As mentioned earlier, the capacity for water uptake through the cut base is small so that avoidance of water stress depends mainly on controlling water loss. In some subjects, stomatal control alone may be sufficient to avoid serious water deficits but for most subjects an environment which discourages transpiration is essential for survival. Transpiration is a diffusion process that is driven by the difference in water vapour partial pressure between the air inside the leaf and the free air around the cuttings, referred to as the leaf to air vapour pressure difference (LAVPD).

Raising the humidity around the cuttings therefore tends to suppress transpiration, but it is rarely possible to prevent transpiration completely in this way, even if a relative humidity of 100% is achieved. This may seem surprising at first, but the reason is as follows. The vapour pressure at the evaporating surfaces inside the leaf can be assumed to be equal to the saturated vapour pressure at the temperature of the leaf. The saturated vapour pressure increases steeply with temperature so that a small elevation of leaf temperature above that of the surrounding air is enough to maintain a positive value of LAVPD. Absorption of radiant energy tends to raise leaf temperature so that only in darkness can saturating the atmosphere be expected to prevent transpiration completely. For the same reason, reducing irradiance helps to restrict transpiration under all conditions.

Evaporation is an energy-absorbing process (the latent heat of evaporation), and transpiration is often limited by the availability of energy. When foliage is kept wet by mist or fog, evaporation of external water absorbs some of the energy that would otherwise be available for evaporation of internal water, with the result that transpiration rates are reduced. A frequently used alternative way of envisaging this process is that evaporation of external water results in evaporative cooling of the leaf, thereby reducing LAVPD and hence transpiration. Wetting of foliage may also benefit the water status of cuttings by absorption of the applied water through the leaf surface, but available data suggest that the amount of water absorbed in this way is minimal (e.g. Grange and Loach, 1983).

From this brief explanation, it is clear that minimizing transpiration from cuttings depends on a combination of low irradiance, high humidity and leaf wetting. To this could be added a number of relatively minor factors including avoidance of vigorous air movement to maintain a high resistance to transport of water vapour across the boundary layer. The integration of these factors has not yet been modelled satisfactorily. Loach (1988a) derived a 'water stress index' by integrating estimates of LAVPD over 24-h periods. However, the model depended on empirical relationships to estimate leaf temperature from irradiance, and incorporated visual estimates of the proportion of the leaf surface covered with water droplets in an attempt to allow for the effect of wetting. Whilst useful as a means of interpreting the observed rooting responses to variation in conditions along a fogged tunnel, the model could not be expected to apply to other situations without modification.

A suitable mechanistic model is not available but would be of considerable value. Such a model would make it possible to carry out a sensitivity analysis on propagation environments and to predict, for example, the extent to which extra leaf wetting could compensate for a given reduction in humidity or increase in irradiance.

The advent of automatic misting equipment reduced the dependence on shade as a means of avoiding water stress, allowing higher irradiance levels to be used (Hess and Snyder, 1955) and greatly extending the range of plants that could be propagated from leafy cuttings. Fogging technology, by making it easier to maintain high humidity, provides a means of avoiding stress more effectively or of increasing irradiance further. As such, it offers new opportunities to propagate difficult plants.

Both the physical principles and practical experience draw attention to the importance of the interaction between irradiance and moisture and point to the need for an effective method of studying these factors in combination. This need led us to develop a controlled propagation environment (CPE) facility specifically for this purpose, incorporating gradients of irradiance and wetting, which is described in detail below.

Different species and varieties exhibit the complete range of rooting ability, from those impossible to root with current knowledge and techniques, to those capable of producing aerial roots on stems before cuttings are collected. One of the objectives of the work described here was to use variation in rooting potential to understand the response of cuttings to components of the aerial environment, and so identify those conditions which are critical for the successful propagation of hitherto difficult-to-root varieties.

Manipulating Rooting Potential Before Cutting Collection

It is helpful within the context of the work described here to investigate interactions between rooting potential and rooting environments within the same genotype. There is opportunity to manipulate rooting potential, especially when dedicated stockplants are used for cutting production, and nine different methods incorporating a wide range of techniques have been used (Howard, 1994). The method chosen here was to grow shoots in darkness for approximately 2 weeks from bud-burst, and then wean them into at least 50% light over a similar period before collecting them as cuttings. A subsidiary method involved

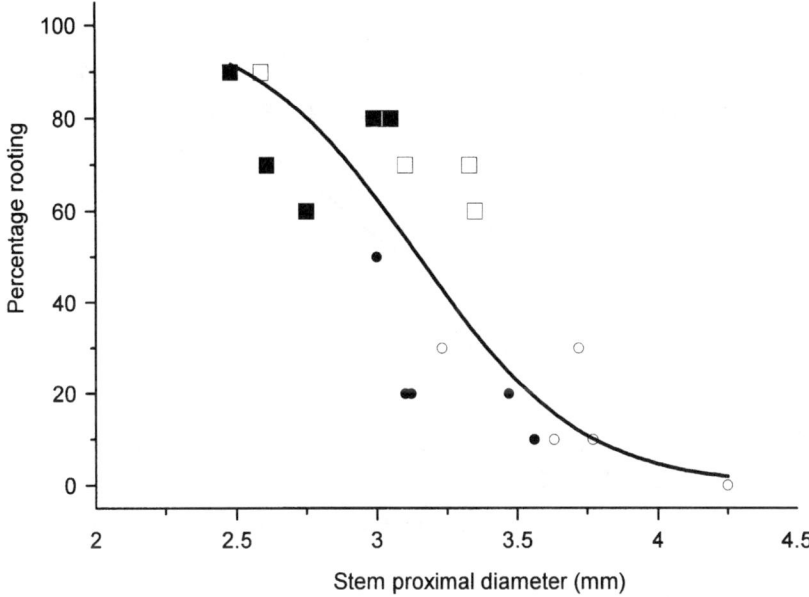

Fig. 7.1. Logistic regression of percentage rooting on stem proximal diameter at the start of propagation in *Syringa vulgaris* 'Madame Lemoine'. Control, ○; dark-pretreated, □; defoliation-pretreated, ●; dark + defoliation-pretreated, ■.

defoliating shoots for 2 weeks before allowing terminal leaves to develop prior to cutting collection.

The effect of dark pretreatment on shoot morphology in, for example, the French hybrid lilac *Syringa vulgaris* 'Madame Lemoine' is to decrease stem mass in terms of fresh and dry weight, dry matter content and stem diameter, without affecting leaf area. Normal leaf photosynthetic and stem respirational rates are not changed by dark-preconditioning, but the total respirational load is reduced because of the smaller stem mass.

There is a strong negative correlation between stem diameter (or weight) and rooting (Fig. 7.1) which appears to have general application in this type of difficult-to-root subject, and which can be predicted from a model incorporating leaf area, leaf number and stem diameter or weight.

Enhanced rooting in dark-preconditioned cuttings is attributed, at least partly, to a favourable distribution of carbohydrate to the potential rooting zone in the 10 days or so before the first root emerges (Fig. 7.2), presumably because carbohydrate is surplus to that required for the respiratory maintenance of the smaller stem, compared with normal light-grown cuttings. The dark-preconditioning response is described fully elsewhere (Howard and Ridout, 1992).

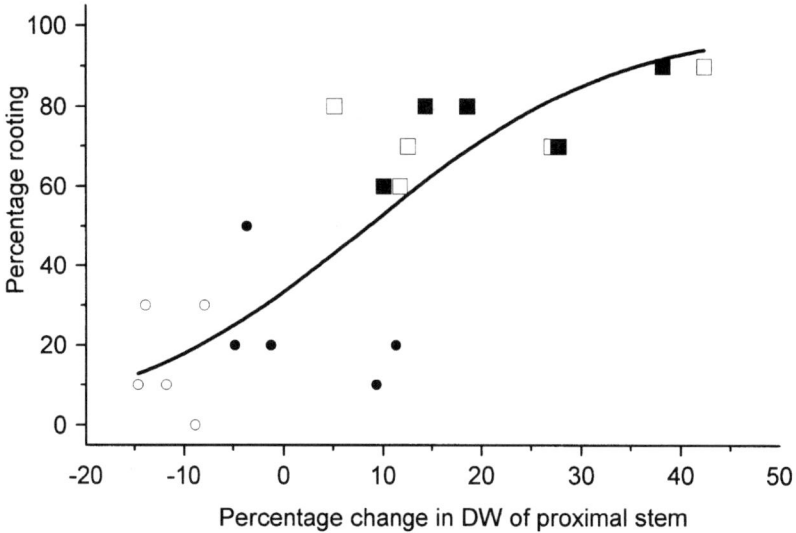

Fig. 7.2. Logistic regression of rooting on the percentage dry weight change of the proximal 35-mm stem segment during the first 10 days in *Syringa vulgaris* 'Madame Lemoine'. Control, ○; dark-pretreated, □; defoliation-pretreated, ●; dark + defoliation-pretreated, ■.

Controlled Propagation Environment Laboratory

Purpose

Much previous research, using conventional propagation facilities, has demonstrated the practical benefits of using different propagation systems such as polyethylene tents, mist and fog and has gone some way to quantifying the separate benefits of leaf wetting, elevated humidity and shading on the water status and rooting of cuttings (e.g. Grange and Loach, 1983; Harrison-Murray et al., 1988). However, the conditions in such facilities are constantly changing in response to variable cloud cover and other meteorological variables, and, in the longer term, in response to diurnal and seasonal cycles. Under these conditions, it difficult to define experimental environments precisely and impossible to reproduce them exactly. As a result, there is little opportunity to replicate physiological measurements in time. For example, the water and turgor potentials of tissues fluctuate hourly (Loach, 1988b), limiting opportunities to quantify relationships between water status, photosynthesis and rooting, even in relation to simple treatment comparisons. These difficulties represent an overwhelming barrier to investigation of interactions between environmental factors, such as the crucial interaction between irradiance and moisture.

By the use of a computer-controlled glasshouse, equipped with multiple layers of shade and supplementary artificial lighting, variation in irradiance and temperature can be reduced but it cannot be eliminated, and such facilities are very expensive, especially on a research scale. These are familiar problems, and the conventional solution is to resort to a fully controlled environment cabinet or room. However, conventional controlled environment facilities are not designed to maintain the unusual conditions required to support leafy stem cuttings prior to rooting. For example, fans provide vigorous circulation of air to minimize local variations in temperature and humidity but such air movement also stimulates water loss and therefore increases the danger of water stress in cuttings. Furthermore, for work on difficult-to-root cuttings, many of which are highly sensitive to water stress, the need for relative humidities close to 100% and frequent spraying with water to keep the foliage wet, would be inimical to the control systems of most cabinets.

These considerations led us to identify a need for a novel form of controlled environment facility that would create a range of the sort of conditions required for rooting cuttings and in a highly reproducible way. To be of practical value, it was also essential that the responses of the cuttings should not be seriously altered by the artificiality of the controlled environment (e.g. the absence of any variation in irradiance throughout the 'day'). Similarly, it was essential to establish a basis for relating conditions in the controlled environment to the wide range of different types of environment used in commercial propagation.

The CPE laboratory was designed to meet these objectives.

Design and construction

The principle adopted was to create a stable background environment in a large controlled temperature room within which propagation environment 'chambers' could be constructed using transparent polyethylene sheeting, artificial illumination being provided by lamps above the chambers.

The dimensions of the temperature-controlled room were 11.9 m × 6.6 m × 4 m high. The walls were constructed of 100-mm thick insulation panels consisting of a core of expanded polystyrene sandwiched between 0.55-mm thick steel sheets coated on the outside with polyvinyl chloride. The ceiling was formed from similar panels, but 150 mm thick, while the floor consisted of 150 mm of solid concrete. It formed the major part of a purpose-designed building based on cost-effective technology used widely in the construction of large cold stores.

A framework of galvanized steel trunking within the room was used to support the lamps and tensioned nylon wires from which the polyethylene chambers were supported. This arrangement provides a high degree of flexibility in the choice of shape and size of the chambers and the positioning of lamps.

Arrangements for current environmental studies

The interaction between the effects of light and moisture were studied within a single chamber by setting up gradients of light and wetting at right angles to each other (Figs 7.3 and 7.4). The gradients were steep so that a very wide range of conditions was created. This chamber is referred to as the gradient CPE (G-

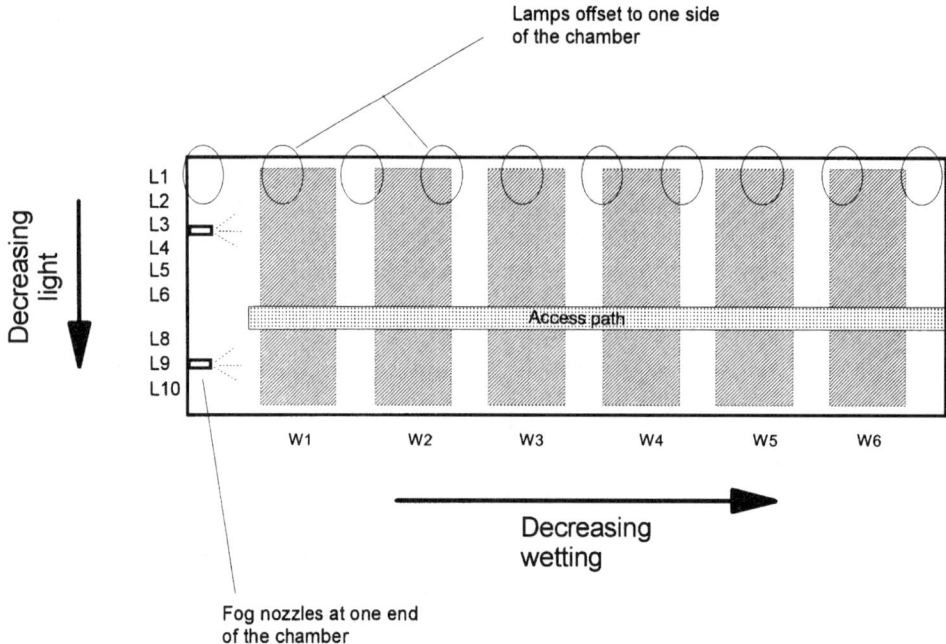

Fig. 7.3. Plan of the gradient controlled propagation environment (G-CPE). The overall dimensions of the chamber are 10 × 2 m.

CPE). How these conditions relate to a normal propagation unit is considered below. The light source was a bank of high-pressure sodium lamps (Type SON-T plus, 400 W, in Philips SGR 140 luminaires) arranged in a line above one side of the chamber. Carefully positioned reflective curtains helped concentrate the light on that side so as to achieve about a tenfold variation in light level across the 2-m wide chamber. The photoperiod was 12 h.

Along the 10-m length of the chamber, a wetting gradient was achieved by injecting fog from two nozzles at one end ('Sonicore' nozzles, using compressed air at approx. 450 kPa). The nozzles operated for 20 s per 50 s when the lights were on and for 30 s per 900 s during the dark period. A 100-mm layer of fine sand over a concrete floor provided drainage for the heavily wetted area and capillary water supply to cuttings at the dry end. The fog was dense at the wet end and just visible at the dry end, where the humidity was usually between 95 and 100%.

For practical purposes, the continuous gradients of light and wetting were considered to be subdivided into discrete levels, designated L1 (highest) to L10 (lowest) for light, and W1 (wettest) to W6 (driest) for wetting (Fig. 7.4). No cuttings could be placed in L7 as this zone provided an access path. The nine levels of light combined with six levels of wetting created a matrix of 54 different environments.

The chamber itself consisted of a clear polyethylene tent, 2.2 m high to a central ridge, the slope of the roof being enough to minimize drips which would

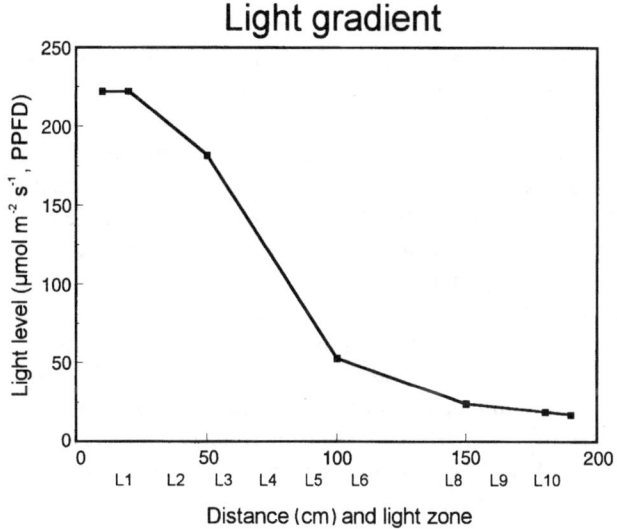

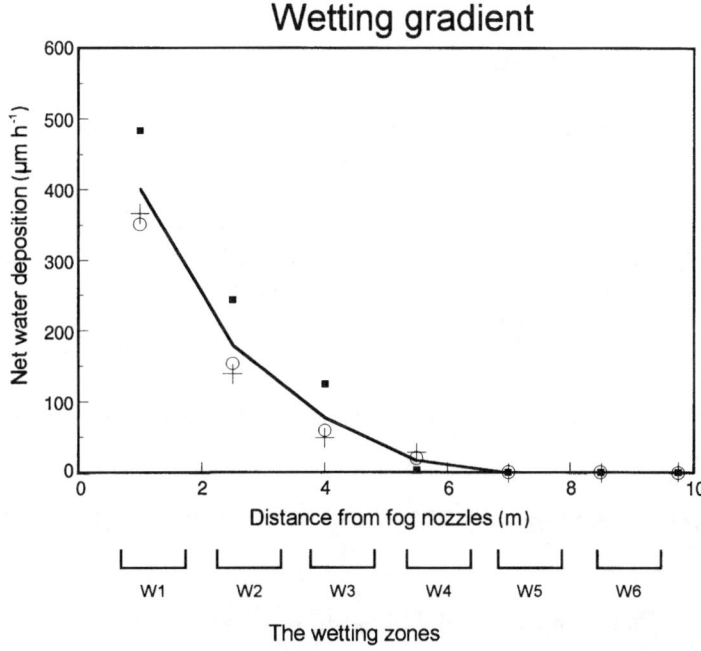

Fig. 7.4. Gradients of light (photosynthetic photon flux density) and wetting in the G-CPE chamber. On the wetting graph, the points indicate the values obtained at three positions across the gradient.

otherwise disturb the wetting gradient. Access was provided from one end through overlapping polyethylene flaps.

The temperature outside the chamber was controlled to 20 ± 2°C. Because of the limited air movement within the chamber, the temperature around the cuttings rose to an average of 23°C when the lights were on, falling back to 20°C during the dark period. There were also temperature differences of up to 2°C between the lowest and the highest light level.

The data in Fig. 7.4 provide a basis for comparison with daily average values which have been observed previously in conventional propagation facilities. Water deposition at the wet end (W1) was equivalent to a conventional mist system, adjusted to keep leaves generously wet at all times, while W2 was equivalent to generous wetting in a fog house in which the high humidity reduces the rate of evaporation of the applied water.

In terms of daily radiation integrals, the maximum photosynthetic photon flux density (PPFD) of 220 µmol m^{-2} s^{-1} is slightly greater than would be received during fine summer weather in a typical propagation house, in which irradiance is reduced to about 20% of that outside by shading. By comparison, during periods of poor weather, 5-day averages can drop to about 70 µmol m^{-2} s^{-1}. Short-term values can be more extreme, but are probably less relevant to the behaviour of cuttings. At other times of year, these figures are reduced by the shorter daylengths, unless shade is reduced to compensate.

'Environmental fingerprints'

The G-CPE system has been used to compare the environmental responses of a wide range of mainly woody ornamental species. To identify approximately the optimum conditions for each subject, four cuttings in each of the 54 different combinations of light and wetting referred to above was usually sufficient replication. Three dimensional graphs, referred to as 'environmental fingerprints', proved the most effective method of analysing and presenting the results (e.g. Fig. 7.5). Where necessary, significance tests were applied by treating groups of six adjacent locations as replicates of larger zones in a 3 × 3 matrix. Where trends were not obvious, multiple regression techniques were used to fit response surfaces to the data, based on third order polynomial equations.

Investigations Using Cuttings with Different Rooting Potential in CPE Conditions

Comparisons between easy- and difficult-to-root genotypes

Cuttings of *Cotinus coggygria* 'Royal Purple' are sufficiently difficult to root that few commercial nurseries can achieve high success rates despite the demand for good-quality specimens of this attractive, red-leaved shrub. Figure 7.5 shows that the frequency of rooting was very sensitive to the aerial environment. Where high irradiance was combined with heavy water deposition, virtually all cuttings rooted (i.e. L1–L4 combined with W1 or W2). In contrast, rooting was prevented completely by high irradiance if cuttings were not heavily wetted (i.e. L1–L4 combined with W4–W6). The data therefore demonstrate that the interaction

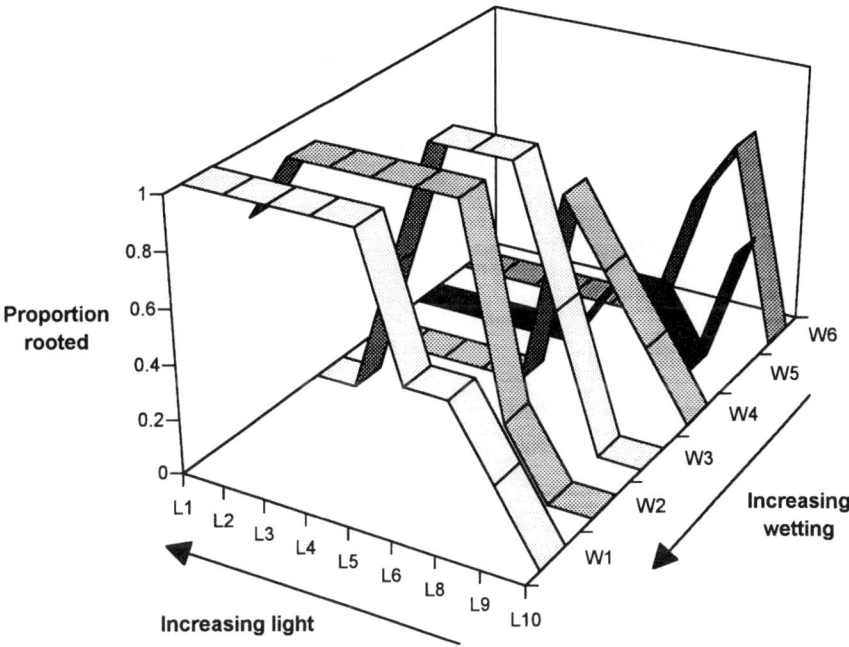

Fig. 7.5. The response of *Cotinus coggygria* 'Royal Purple' cuttings to the range of conditions in the G-CPE. (To convert 'proportion rooted' into 'percentage rooted', multiply by 100; the former is used here and elsewhere because of the small number of cuttings at each of the 54 locations, on which the plotted values are based.)

between the effects of light and moisture on cuttings can be very strong. Indeed, the strength of the interaction is such that it should be taken into account in all comparisons of rooting environments. For example, a comparison of a mist system with a fog system would give very different results depending on the degree of shading applied, and would be virtually meaningless if the degree of shading was not the same in both systems. Similarly, the weather prevailing at the time of the experiment could influence the outcome.

However, the data also provide evidence for an effect of irradiance that is independent of moisture. Irrespective of the amount of water deposited on the cuttings, few cuttings rooted at very low irradiance (from L8 to L10, equivalent to a PPFD of less than 40 µmol m^{-2} s^{-1}). Under these conditions, a decrease in the mean dry weight of the stems of the cuttings was observed, whereas stem dry weight increased at higher irradiance.

Cryptomeria japonica 'Elegans Compacta' is a conifer with narrow needle-like leaves and is generally considered much more readily rooted than *C. coggygria* 'Royal Purple'. The cuttings of *C. japonica* 'Elegans Compacta' responded to environment in the opposite way to those of *C. coggygria* 'Royal Purple' (Fig. 7.6). Irrespective of light level, heavy water deposition suppressed rooting almost completely, whereas most cuttings rooted under the combination of high light with minimal wetting, conditions that prevented rooting in *C. coggygria* 'Royal

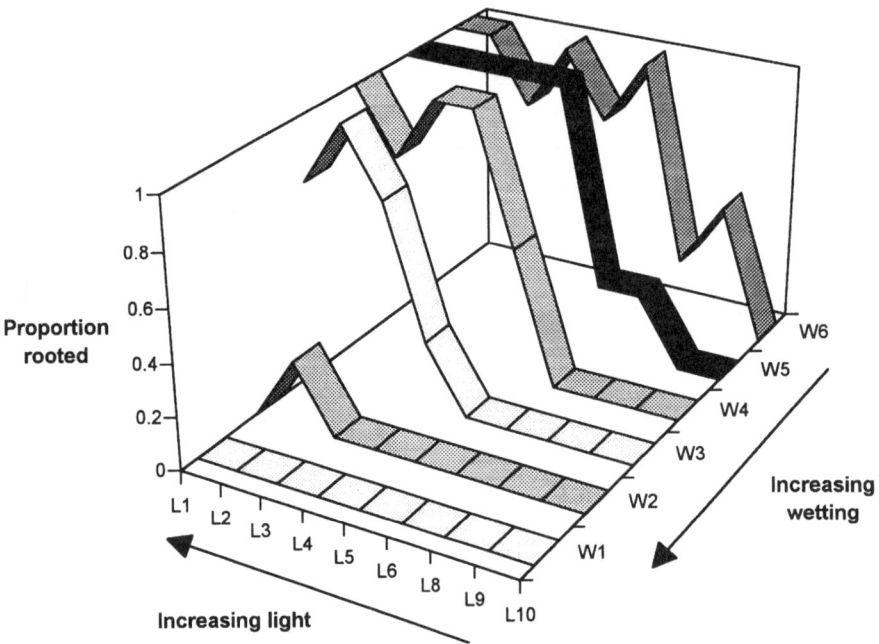

Fig. 7.6. The response of *Cryptomeria japonica* 'Elegans Compacta' cuttings to the range of conditions in the G-CPE.

Purple'. As the amount of wetting decreased, the minimum irradiance at which any of the cuttings rooted also decreased progressively. However, even at the dry end of the chamber (W5 and W6), the proportion of cuttings which rooted dropped sharply when PPFD was less than 40 µmol m^{-2} s^{-1} (i.e. L8–L10). This is similar to the threshold irradiance observed in *C. coggygria* 'Royal Purple' cuttings. This similarity suggests that the response of rooting to irradiance *per se*, evident when other conditions are non-limiting, is rather similar in these two species.

Another common feature of the response surfaces of these two species is that the diagonal running from high light–heavy wetting to low light–low wetting coincides approximately with a line of sharp change in rooting percentage. The difference between the species is the direction of the sharp change that occurs along this diagonal. The tendency for similar rooting percentages to be observed along the diagonal, and lines running parallel to it, implies that light tended to counteract the effect of wetting, at least once irradiance was above the threshold referred to above. This applied equally to the beneficial effect of wetting on *C. coggygria* 'Royal Purple' cuttings and its adverse effect on *C. japonica* 'Elegans Compacta' cuttings. This result is consistent with the expected stimulation of transpiration by light. Stimulation of transpiration is expected partly because it tends to raise leaf temperature and thus LAVPD, the driving force for transpiration, and partly because stomata tend to open as irradiance increases (Warrit *et al.*, 1980).

The adverse effect of wetting on *C. japonica* 'Elegans Compacta' is therefore more likely to be due to excessive suppression of transpiration than to a deleterious effect of leaf wetting *per se*. It remains to be determined how a low rate of transpiration can inhibit rooting. It is possible that some degree of water stress is required to trigger the root initiation process in this subject but it is also possible that a certain amount of transpiration is required to prevent damaging overhydration of the tissues close to the basal wound, which might include intercellular air spaces becoming filled with water.

Why the two species show such contrasting responses to wetting is the subject of current research. Measurements of water uptake by freshly collected cuttings, using an automatic potometer system, showed tenfold greater rates of uptake per cutting in *C. coggygria* 'Royal Purple' than in *C. japonica* 'Elegans Compacta', but when the data were expressed per unit leaf area the difference was only twofold. Furthermore, reducing the leaf area of *C. coggygria* 'Royal Purple' cuttings by 75% did not alter their response to wetting substantially.

Nurserymen consider *C. japonica* 'Elegans Compacta' cuttings much easier to root than cuttings of *C. coggygria* 'Royal Purple'. However, the results from the G-CPE show that the rooting potential of both cultivars is high but that they have very different environmental requirements. The reason that *C. coggygria* 'Royal Purple' is considered much more difficult to root is that few commercial propagation systems provide the combination of conditions that allow its full rooting potential to be expressed. The practical means of achieving these conditions are considered further in a later section.

Forsythia × *intermedia* 'Lynwood' is recognized as a cultivar that roots extremely readily from leafy cuttings. With this subject, 100% of cuttings rooted over the full range of conditions provided in the G-CPE. The effects of environment were confined to the number and length of roots per rooted cutting, which were greatest where high irradiance was combined with heavy wetting (data not shown). Similar results were obtained with *Weigela florida* 'Variegata'. Amongst the factors that confer such a wide environmental tolerance on these subjects, evidence from carbon dioxide exchange and from changes in dry weight indicated that leaves of the two easy-to-root subjects have a lower light compensation point than leaves of *C. coggygria* 'Royal Purple'. When high irradiance was combined with minimal wetting, cuttings of the easy-to-root subjects wilted less severely than cuttings of *C. coggygria* 'Royal Purple', despite similar leaf area per cutting. Porometer measurements of stomatal conductance indicated that this was attributable to more rapid closure of stomata in response to declining leaf water status. This may explain how cuttings of the easy-to-root subjects were able to root under conditions of such high transpiration demand. However, it may also be relevant that roots emerge about 10 days earlier than in *C. coggygria* 'Royal Purple' so that the tissues are under water stress for a substantially shorter period.

Comparisons within the same genotype

Rooting potential also varies within a single genotype and can be manipulated by various pretreatments. One of the most powerful is to exclude light temporarily from the stockplants so that the shoots that will be taken as cuttings initially develop in darkness. In exploring the possible involvement of carbon balance in this response, the environmental response of dark- and light-grown cuttings was compared using the G-CPE approach.

Terminal cuttings of *S. vulgaris* 'Madame Lemoine', approximately 20 cm long and with one pair of fully expanded leaves, together with smaller distal leaves, were prepared from light-grown and dark-pretreated shoots. They were placed in an earlier version of the G-CPE facility in which PPFD ranged from mean values of 289 to 17 µmol m^{-2} s^{-1} and water deposition ranged from mean values of 209 to 13 µm h^{-1}. In this case the gradients were divided into just four zones, thereby creating a simpler 4 × 4 experimental matrix, with a greater number of replicate cuttings in each part of the matrix. Results were consistent over 2 years and are combined here; more detailed information is given elsewhere (Howard and Harrison-Murray, 1995).

Responses prior to rooting

A subsample of cuttings taken 10 days after insertion showed that the fresh weight of the proximal 35 mm of stem (potential rooting zone) decreased with decreasing wetness at the highest light level, but not at the lowest light level, with a similar trend for both sources of cutting.

The dry weight of the proximal stem section decreased with decreasing light, but was depressed relatively little by the severe wilting which occurred in the high light–low wetting zone.

When data for day 10 were compared with those at the start of the experiment it was found that at the two higher light levels the dry matter content of the proximal stem section increased, especially in the dark-pretreated cutting. The opposite occurred at the two lower light levels, where net losses in dry matter content occurred over this period, which were more severe in the dark-pretreated cuttings than in the normal light-grown ones (Fig. 7.7).

The frequency and extent of stem necrosis caused by the physiological breakdown of tissue generally reflected the loss of dry matter, with rotting being most severe in the dark-pretreated cuttings at the lowest light level.

Callus production and rooting

Callus was prevented where stems developed proximal necrosis, although in rare cases a ring of callus developed above the dead tissue. High light and generous wetting favoured callus production. Water-stressed cuttings in the high light–low wetting zones produced little callus, despite the general absence of necrosis.

No cuttings rooted in the four zones which combined the two highest light levels and the two driest areas, nor in the single darkest and wettest zone (Fig. 7.8). Otherwise, the ability of cuttings to root over a wide range of light and wetting in only 3 weeks was surprising, indicating among other things, that rooting was not dependent on cuttings being held above the light compensation point. Even so, most frequent rooting occurred in the high light zones with generous wetting, conditions which the dark-pretreated cuttings were able to exploit much more than normal light-grown ones. This contrasts with the fact

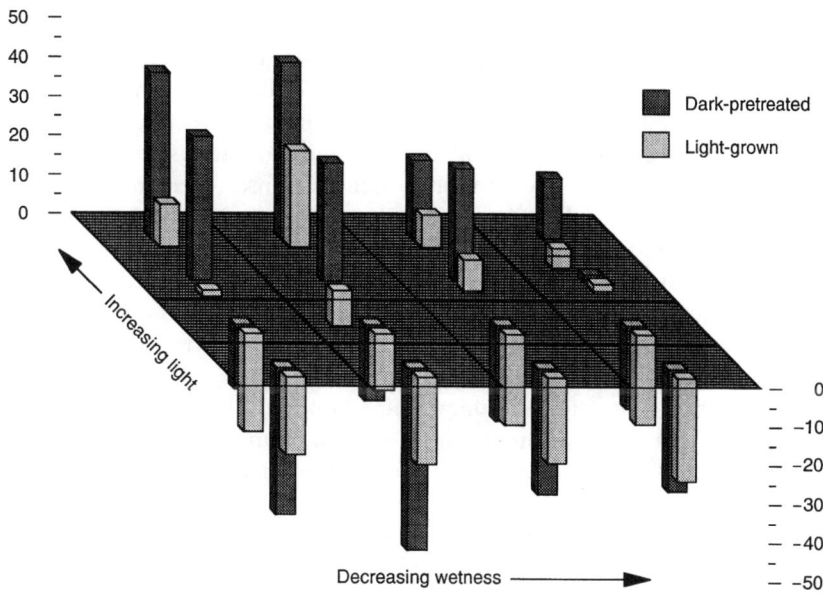

Fig. 7.7. Percent dry weight changes in the proximal 35 mm of stem over the first 10 days in relation to location within a G-CPE (*Syringa vulgaris* 'Madame Lemoine').

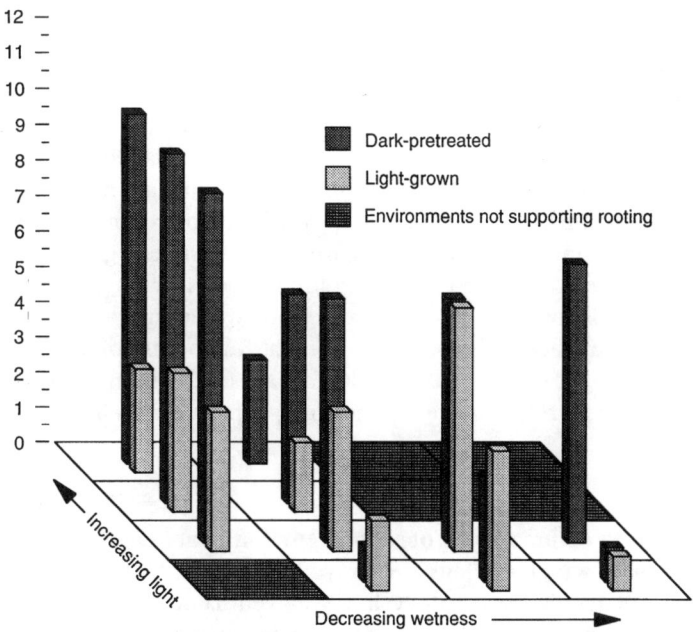

Fig. 7.8. Number of cuttings rooted out of 12 in relation to location within a G-CPE (*Syringa vulgaris* 'Madame Lemoine').

that light-grown cuttings rooted more frequently than dark-pretreated cuttings at the lower light levels because fewer of the latter survived in conditions where they were unable to direct carbohydrate to the base of the stem.

From these results, it would appear that dark preconditioning increases the ability of the *S. vulgaris* 'Madame Lemoine' cuttings to exploit a favourable environment, but does not alter their response to environment substantially. Characteristic of this response, probably in common with that of many subjects which are regarded as difficult to root, is that rooting is prevented by otherwise beneficial high levels of light, unless these are combined with generous leaf wetting to avoid stress. While the benefit of dark preconditioning appears to relate to a strongly positive carbon balance, water stress prevented rooting even though it did not prevent a positive carbohydrate balance.

From a practical standpoint, it is important to note that the wide range of environments from bright and wet to dull and relatively dry that supported rooting of difficult subjects in these experiments are not those normally met with in commercial practice. In nurseries, high light most often is accompanied by lack of adequate wetting, and conditions in heavy shade are often excessively wet, circumstances that led here to no rooting, as was the case also for *C. coggygria* 'Royal Purple' and *C. japonica* 'Elegans Compacta' respectively described earlier.

Opportunities for Improved Environmental Sensing and Control

The results from the G-CPE facility provide an insight into the opportunities to improve rooting of difficult-to-root genotypes by optimizing the rooting environment. They also provide pointers to the improvements in equipment needed to realize these opportunities on commercial nurseries.

The results for *C. coggygria* 'Royal Purple' show that some plants are considered difficult to root because they have a relatively narrow environmental tolerance and the conditions required for rooting are rarely met by the facilities in commercial nurseries. If an appropriate environment is available, close to 100% rooting of cuttings prepared from actively growing shoots is easily achievable. The environment required is one in which the potential transpiration rate is low enough to avoid serious water stress despite moderately high irradiance. Such conditions are optimal for rooting of a wide range of shrubs especially those with relatively large leaves. Why then is such a combination of conditions rarely achieved in existing propagation facilities?

Mist systems

The most common type of rooting environment in use commercially is the mist system in which cuttings are sprayed intermittently with fine droplets of water. They generally are operated in a well-ventilated house so that the humidity is rarely very high. Although leaves may be well wetted, the low humidities increase potential transpiration compared with the heavily wetted zone in the G-CPE so that cuttings of *C. coggygria* 'Royal Purple' suffer water stress. Under these marginal conditions, irradiance can determine whether stress is severe enough to prevent rooting. The prevailing weather and the amount of shade applied are therefore critical, with the consequence that rooting percentage is very variable from one propagation to another and also between different nurseries.

Fog systems

Fog systems employ more finely atomized water which remains suspended in the air for at least a few minutes. Where fog is used in commercial nurseries, the intention is usually to raise the humidity around the cuttings without wetting them heavily. In the absence of heavy wetting, potential transpiration is again too high for successful rooting of *C. coggygria* 'Royal Purple'. Another problem that afflicts all systems in which ventilation is minimized to retain humidity, is that heat is also retained, often leading to air temperatures in excess of 40°C. If the response to this problem is to apply extra shading, rooting may be impaired by low irradiance. Alternatively, if ventilation is provided when the temperature is considered dangerously high, then the aim of avoiding water stress is unlikely to be met.

Various factors, such as the use of reflective shade and the size and shape of the house, influence the severity of the heat retention problem. However, the most reliable means of preventing excessive temperatures is to ensure that the humidification system can maintain the desired high humidity in the face of as much ventilation as is required to remove excess heat. In this respect, fog has the great advantage that it presents an enormous evaporative surface, and can continue to evaporate as the temperature of the air in which it is suspended rises. Furthermore, evaporative cooling associated with evaporation of the fog droplets greatly reduces the amount of ventilation required. Unfortunately, the majority of the fog systems which have been installed in nurseries in the UK are not capable of a sufficiently large output to exploit this advantage.

Measuring evaporative demand

Defining the optimal environment is made more difficult by the strong interactions between environmental factors such as that between irradiance and wetting which emerged clearly in the results from the G-CPE. That particular interaction is almost certainly an example of many that arise because cuttings are highly sensitive to the evaporative demand of the environment which itself depends on the combined effects of many factors. Commercially available sensors that are used to control the operation of fog or mist nozzles respond principally to just one of the relevant factors: wetting, humidity or light. With such sensors, optimum settings will depend on the levels of the other factors which are not being sensed. This makes manual adjustments necessary to compensate for changes in these other factors due to variation in the weather, shading or ventilation. It also hinders the transfer of information about optimal conditions for the propagation of particular plants.

Previous approaches

Loach (1988a) explored the use of estimates of LAVPD as an integrated measure of potential transpiration which he referred to as a 'stress index'. This requires a knowledge of leaf temperature which is technically difficult to measure accurately or to estimate from other variables. It also shares with any approach based on a mathematical model the difficulty of monitoring the amount of water that is being deposited.

A novel sensor

A more promising approach is based on a physical model of a transpiring leaf from which an electrical signal can be derived that is proportional to the rate of evaporation from the model. The principle of this approach was established by Harrison-Murray (1991) and this has since been used to develop a working 'evapo-sensor'. The evapo-sensor has been used to control both fog and mist systems successfully. Using commercially available interfacing equipment, it is possible to adjust the target evaporative demand to meet the needs of particular subjects. It is found that, compared with a humidity sensor for fog, or a wetness sensor for mist, the use of an evapo-sensor causes greater changes in the output of mist or fog in response to changing weather conditions, particularly changes in irradiance. In a fog system under evapo-sensor control, little or no fog is injected during dull weather and the surface of the cuttings may become dry, but during hot sunny weather a dense fog will form, wetting the cuttings heavily as well as maintaining a saturated atmosphere. This regime minimizes water stress during sunny weather whilst avoiding a number of practical problems associated with overwetting. Combined with shade to reduce light levels to about 20% of available light, it also leads to a consistently high rooting percentage of difficult subjects such as *C. coggygria* 'Royal Purple'.

Once cuttings have rooted, it is necessary to acclimatize, or 'wean', them to the more desiccating environment in which they will be grown subsequently. Using a control system based on an evapo-sensor, it is possible progressively to increase the maximum evaporative demand to which the cuttings are exposed, a regime which has proved ideal for weaning difficult subjects.

When used to monitor evaporative demand outside, the integrated output of the evapo-sensor correlated with actual water use of container-grown shrubs better than established methods of estimating potential evapotranspiration (e.g. pan evaporimeter and the Penman estimate). It is possible, therefore, that the evapo-sensor may have application in the control of irrigation. Opportunities for the commercial production and marketing of this sensor are being evaluated.

Conclusions and Further Opportunities

The ability to avoid excessive water stress appears to be the critical factor determining rooting in leafy softwood cuttings, but critical stress levels are difficult to quantify because they differ with variety and species, and at different stages of propagation.

In addition to rooting quickly, species defined as easy to root may be programmed genetically to conserve water more effectively than difficult-to-root species. Success with the latter will depend largely on balancing the need for relatively high light (approx. 20% of available light during summer in the UK) with generous leaf wetting and high humidity.

This focuses attention on understanding how water stress negates rooting. Current questions include whether rapid and essential cell division is prevented, as suggested by the greatly reduced callus development shown by *Syringa* cuttings in high light–low wetting conditions, or whether, for example, the translocation from the leaf of essential root promoting factors, other than carbohydrate, is negated.

The ability to obtain a better understanding of the role of water stress in preventing adventitious rooting, even in dark-pretreated cuttings with enhanced rooting potential, has important practical consequences.

Other work in this programme has shown that rooting can occur relatively freely in the otherwise difficult to root *Syringa* cuttings in a highly localized area of stem behind the apex. This was demonstrated experimentally by the induction of aerial roots following the daily application of aqueous solutions of the potassium salt of IBA or naphthalene acetic acid to cuttings held in a favourable rooting environment.

Any attempt to exploit this response by propagating very small immature apical cuttings will carry with it the need for very effective environmental control.

At the other extreme, it is often difficult after propagation to induce the branch framework necessary for the production of high-quality shrubs such as *C. coggygria* 'Royal Purple', because small newly rooted plants lack the necessary vigour to branch well in response to pruning. However, the required branches form readily on the vigorous stockplant in response to shoot tipping, leading to the need to be able to propagate particularly large leafy cuttings with multiple branches already present.

Acknowledgements

This work, and the construction of the Controlled Propagation Environment Laboratory, was funded by the Ministry of Agriculture, Fisheries and Food. Figures 7.1, 7.2, 7.7 and 7.8 are published by permission of the *Journal of Horticultural Science*.

References

Davies, F.T., Jr, Davis, T.D. and Kester, D.E. (1994) Commercial importance of adventitious rooting to horticulture. In: Davis, T.D. and Haissig, B.E. (eds), *Biology of Adventitious Root Formation*. Plenum Press, New York, pp. 53–59.

Davis, T.D. (1988) Photosynthesis during adventitious rooting. In: Davis, T.D., Haissig, B.E. and Sankhla, N. (eds), *Adventitious Root Formation in Cuttings*. Dioscorides Press, Portland, Oregon, pp. 79–87.

Grange, R.I. and Loach, K. (1983) The water economy of unrooted leafy cuttings. *Journal of Horticultural Science* 58, 9-17.

Harrison-Murray, R.S. (1991) An electrical sensor for potential transpiration: principle and prototype. *Journal of Horticultural Science* 66, 141-149.

Harrison-Murray, R.S., Howard, B.H. and Thompson, R. (1988) Potential for improved propagation by leafy cuttings through the use of fog. *Acta Horticulturae* 227, 205-210.

Hartmann, H.T., Kester, D.E. and Davies, F.T., Jr (1990) Techniques of propagation by cuttings. In: *Plant Propagation; Principles and Practices*, 5th Edn. Prentice-Hall, Englewood Cliffs, New Jersey, pp. 256-304.

Hess, C.E. and Snyder, W.E. (1955) A physiological comparison of the use of mist with other propagation procedures used in rooting cuttings. In: *Proceedings of the 14th International Horticultural Congress*, Vol. 2, pp. 1125-1139.

Howard, B.H. (1965) Regeneration of the hop plant (*Humulus lupulus* L.) from softwood cuttings. I. The cutting and its rooting environment. *Journal of Horticultural Science* 40, 181-191.

Howard, B.H. (1994) Manipulating rooting potential in stockplants before collecting cuttings. In: Davis, T.D. and Haissig, B.E. (eds), *Biology of Adventitious Root Formation*. Plenum Press, New York, pp. 123-142.

Howard, B.H. and Harrison-Murray, R.S. (1995) Responses of dark-preconditioned and normal light-grown cuttings of *Syringa vulgaris* 'Madame Lemoine' to light and wetness gradients in the propagation environment. *Journal of Horticultural Science* 70, 989-1001.

Howard, B.H. and Ridout, M.S. (1992) A mechanism to explain increased rooting in leafy cuttings of *Syringa vulgaris* 'Madame Lemoine' following dark-treatment of the stockplant. *Journal of Horticultural Science* 67, 103-114.

Loach, K. (1988a) Characterisation of optimal environments for rooting leafy cuttings. *Acta Horticulturae* 226, 403-412.

Loach, K. (1988b) Water relations and adventitious rooting. In: Davis, T.D., Haissig, B.E. and Sankhla, N. (eds), *Adventitious Root Formation in Cuttings*. Dioscorides Press, Portland, Oregon, pp. 29-46.

Macdonald, B. (1986) Softwood cuttings. In: *Practical Woody Plant Propagation for Nursery Growers*. Timber Press, Portland, Oregon, pp. 289-297.

Thomas, A. and Harrison-Murray, R.S. (1995) What makes peat so good for growing. *Grower* 23, 21-22.

Veierskov, B. (1988). Relations between carbohydrates and adventitious rooting. In: Davis, T.D., Haissig, B.E. and Sankhla, N. (eds) *Adventitious Root Formation in Cuttings*. Dioscorides Press, Portland, Oregon, pp. 70-78.

Warrit, B., Landsberg, J.J. and Thorpe, M.R. (1980) Responses of apple leaf stomata to environmental factors. *Plant, Cell and Environment* 3, 13-22.

8 The Use of Mutants and Molecular Biology to Understand Competence for Root Formation

W.P. HACKETT, S.T. LUND, A.G. SMITH AND N.E. OLSZEWSKI

Departments of Horticultural Science and Plant Biology, University of Minnesota, St Paul, MN 55108, USA

Introduction

Many research approaches have been used in attempts to explain the basis for differences in competence for root initiation among species, clones and developmental phases within clones. Competence is defined as the ability to respond to a signal for a specific morphogenetic process, in this case, adventitious root initiation. For root initiation, auxin is usually considered to be the signal or one of the signals. Because of the long reproductive cycle in most woody perennial species, in which competence for root initiation is a limiting and practically important variable, a genetic approach has not been feasible for them. With the rather recent development of molecular biological techniques, methods of generating and screening for monogenic mutants, and methods of tagging and cloning affected genes, a genetic approach to understanding the basis of competence may now be the method of choice. Because rooting is likely to be a complex, multistep process (Malamy, and Benfey, 1997), a series of single gene mutants with developmental blocks in various steps in the process would provide a powerful tool for understanding competence for root initiation.

Mutants for Root Initiation

A few monogenic mutants for altered root initiation have been identified (Table 8.1). They can be categorized into three general groups: (i) those which have an excess of lateral and adventitious roots (Boerjan *et al.*, 1995; Celenza *et al.*, 1995; King *et al.*, 1995; N. Olszewski, unpublished); (ii) those in which auxin does not induce or has a small effect on lateral and adventitious root initiation (Celenza *et al.*, 1995; Cheng *et al.*, 1995; Lund *et al.*, 1996); and (iii) those in which there is aberrant development of lateral root initials (Celenza *et al.*, 1995; Cheng *et al.*, 1995). For those in which there is an excess number of roots, it has been shown

Table 8.1. Monogenic mutants for root initiation and development that may be useful for analysing the basis of competence for adventitious root initiation.

Species	Mutant name	Mutation type	Root phenotype	Root response to auxin	Endogenous auxin levels	Gene tagged or mapped	Gene cloned	Citation
Arabidopsis thaliana	superroot, sur 1-1 to 1-7	Recessive	Excess lateral roots; roots on hypocotyl	Excised roots are auxin autotrophic	Elevated	Mapped to chromosome 2	No	Boerjan et al. (1995)
Arabidopsis thaliana	rooty, rty-1 to 4	Recessive	Excess lateral roots; roots on hypocotyl	Excised roots are auxin autotrophic	Elevated	Mapped to chromosome 2; rty-3 T-DNA tagged	Yes	King et al. (1995); J. Ecker, J. Reed and N. Olszewski, unpublished data.
Arabidopsis thaliana	alf 1-1	Recessive	Excess lateral roots; roots on hypocotyl	Unknown	Unknown	No	No	Celenza et al. (1995)
Arabidopsis thaliana	rml 1	Recessive	Primary and lateral roots cease growth	Adventitious roots induced by auxin, root growth not rescued by auxin	Unknown	Mapped to chromosome 4	No	Cheng et al. (1995)
Arabidopsis thaliana	alf 4-1	Recessive	No lateral roots	Adventitious roots not induced by auxin	Unknown	No	No	Celenza et al. (1995)
Arabidopsis thaliana	rml 2	Recessive	No lateral roots, primary roots cease growth	Adventitious and lateral roots not induced by auxin		Mapped to chromosome 3	No	Cheng et al. (1995)
Nicotiana tabacum 'Xanthii'	rac	Dominant	Primary roots cease growth	Adventitious roots not induced by auxin	Auxin conjugates normal	No	No	Muller et al. (1985); Lund et al. (1996)

that these mutants have elevated levels of indoleacetic acid (IAA) and its conjugates (Boerjan et al., 1995; King et al., 1995). The affected gene has been tagged and has been cloned in one of these mutants (N. Olszewski, unpublished). One of the mutants which has no response to auxin for root initiation has been proposed to be an auxin sensitivity mutant (Barbier-Brygoo et al., 1990). The nature of the gene involved and the basis of the mutant phenotype have not been determined unequivocally for any of the mutants. The mutants in which auxin treatment does not induce or has a small effect on root initiation may be most useful for analysing the basis for competence of root initiation as it is lack of competence for rooting that is important horticulturally. The *rml* 2 and *alf* 4-1 *Arabidopsis* mutants might be particularly useful. Ernst (1994) listed and discussed characteristics of a model system or species for studying causal mechanisms in adventitious rooting. He concluded that *A. thaliana* was superior in many ways to any other species. All but one of the mutants listed in Table 8.1 are of *A. thaliana*.

We have done research on two of the mutants shown in Table 8.1, *rty*-3 and *rac*. *rty*-3 is representative of a group of *A. thaliana* mutants which have an excess of lateral and adventitious roots, and the *Nicotiana tabacum* 'Xanthii' mutant *rac* is auxin resistant and does not produce adventitious roots in response to exogenous auxin treatment. These two mutants are extreme examples of genotypes with excessive or no ability to form adventitious roots and may, therefore, be useful in the analysis of the basis of high and low rooting competence in plants.

We will first describe briefly what is known about *rty*-3 and other similar mutants, and then give a more detailed discussion of our experimental analysis of the developmental and molecular basis of the lack of rooting competence in the *rac* mutant.

rty-3 and Other Similar Mutants

rty-3 seedlings germinate normally, but form large numbers of adventitious roots on the hypocotyl and cotyledons shortly after germination. This phenotype could be due to higher competence for root initiation or a high level of a signal such as IAA and/or its conjugates. Genetic analysis indicates that the phenotype of *rty*-3 is due to a single recessive mutation and that it is allelic to *rty*-1 (King et al., 1995), invasive *root*-1 and -2 (King et al., 1995), and *alf* 1-1 (Celenza et al., 1995). A phenocopy of *rty*-1 is produced when wild-type seedlings are grown on a medium containing a relatively high level of the synthetic auxin indolebutyric acid (IBA) (King et al., 1995). Because *rty*-1 contains elevated levels of free IAA and IAA conjugates, King et al. (1995) have hypothesized that the *RTY* product is involved in regulating auxin metabolism.

rty-3 was identified in the progeny of plants regenerated after *Agrobacterium tumefaciens* transformation treatment. The *rty*-3 mutant was most probably generated as a result of a T-DNA or GUS reporter gene insertion into the *RTY* gene. We have cloned the T-DNA and some flanking genomic sequences from *rty*-3. The flanking sequences have been used to obtain both cDNA and genomic clones corresponding to this region. Partial sequences have been obtained for the region flanking the T-DNA, the cDNA and the wild-type genomic clones. These sequences indicate that the *rty*-3 T-DNA resides near the 3' end of the open reading frame encoded by the cDNA. Computer searches indicated that the cDNA is most similar to tyrosine aminotransferase from humans and mouse. This result is consistent with the hypothesis that the *RTY* gene is involved in some aspect of auxin metabolism.

Because of the formation of large numbers of adventitious roots on the hypocotyls and cotyledons of the *rty*-3 mutants, the high endogenous IAA levels of an allelic mutant, *rty*-1, and the similarity of the cDNA sequence to a tyrosine aminotransferase gene from humans, we also hypothesize that the *RTY* product is involved in auxin metabolism.

We currently are doing work to determine the mechanism by which mutations in the *RTY* gene affect IAA metabolism. Our working hypothesis is that the *RTY* gene product is, as the sequence predicts, an aminotransferase such as prephenate aminotransferase (Bentley, 1990) which is involved in phenylalanine and/or tyrosine biosynthesis, or some other pathway that competes for an IAA precursor. Because a mutation in *RTY* would be predicted to block the synthesis of these amino acids and thereby cause the accumulation of chorismate, the common shikimate pathway precursor for phenylalanine, tyrosine and tryptophan (Bentley, 1990); the accumulated chorismate could then feed into the tryptophan pathway and contribute to excess IAA synthesis.

From a physiological point of view, the high endogenous IAA levels and the nature of the mutant gene indicate that the *rty* phenotype is due to high amounts of the signal IAA but not high competence for root initiation.

Genetic and Developmental Basis of the Lack of Rooting in the *rac* Mutant

Muller *et al.* (1985) identified an auxin-resistant mutant, *rac*, of tobacco (*N. tabacum* var. Xanthii) which is impaired in primary root development. They generated the mutant genotype via UV irradiation of wild-type mesophyll protoplasts; mutant protoplasts were selected that had the ability to proliferate in media containing levels of naphthaleneacetic acid (NAA) that were toxic to wild-type protoplasts. A mutant line was recovered that confers a tenfold increase in the resistance of cell suspensions to auxin and a co-segregating impaired primary root development phenotype in seedlings. *rac* plantlets from seed or stem cuttings from mutant plantlets do not form adventitious roots. Mutant shoots grafted on to wild-type rootstock are fertile and produce viable seed upon selfing. A karyotype analysis of mutant petal cells revealed that mutant plants are diploid, even though they were derived from mutagenized haploid protoplasts. Progeny tests showed that the auxin resistance and the impaired primary root development phenotypes are dominant and caused by a mutation in a single gene. *rac* plantlets are not impaired in auxin transport and do not contain altered levels of auxin conjugates (Caboche *et al.*, 1987).

Our experiments (Lund *et al.*, 1996) demonstrate that adventitious root formation does not occur when heterozygous and homozygous *rac* microshoot-derived stem cuttings are treated *in vitro* with IBA concentrations ranging from 0.5 to 500 µM. Optimal root initiation was obtained in wild-type cuttings at 5.0-10.0 µM IBA. Histological analysis showed that some phloem parenchyma or inner cortical parenchyma cells in wild-type stem cuttings undergo adventitious root morphogenesis when treated with 5.0 µM IBA. The same cell types in heterozygous and homozygous *rac* stem cuttings undergo mitoses in response to auxin, but never form adventitious root meristems. We conclude that *rac* mutants are incompetent for adventitious root initiation in response to auxin. The lack of adventitious root initiation in *rac* stem cuttings is phenotypically distinct from the aberrant primary root development in *rac* stem seedlings in which case a root meristem is formed but aborts during germination (Lund *et al.*, 1996). The *rac* mutation appears to block an essential process for auxin induction of adventitious root initiation but not cell division in phloem parenchyma or inner cortical parenchyma cells. Comparisons of *rac* heterozygous and homozygous seedling primary root length and callus formation in response to auxin in stem cuttings indicate that *rac* copy number is correlated to the degree of expression of these two phenotypes.

Based on responses of *rac* and wild-type protoplast and cell suspension to auxins, Barbier-Brygoo *et al.* (1990) concluded that expression of the dominant *rac* mutation causes lowered sensitivity to auxin by reducing the number of functional auxin-binding sites at the plasmalemma. However, our results show that formation of adventitious root meristems is completely blocked in *rac* stem cuttings even at high auxin concentrations, while cell division occurs readily in response to auxin. Assuming that Barbier-Brygoo's conclusion is correct, our results suggest that the cellular incompetence of *rac* stem parenchyma cells, specifically for adventitious root initiation but not for cell division, may reflect a higher threshold of auxin binding required for proper signalling for root initiation than for cell division. Alternatively, cellular incompetence of *rac* stem cells for adventitious root initiation but not for cell division may be due to the existence of different auxin receptors for the two responses.

Molecular Analysis of the Basis of Root Initiation Incompetence in the *rac* Mutant

Our developmental analysis indicated that the *rac* mutation blocks adventitious root initiation but not unorganized cell divisions in response to auxin (Lund *et al.*, 1996). This work also suggests that the *rac* mutation blocks adventitious root initiation prior to the first organized cell divisions that normally lead to the formation of an adventitious root meristem. This difference in competence for root initiation but not cell division in wild-type and *rac* cuttings provides an excellent experimental system for analysis of the molecular basis of competence for root initiation (Lund *et al.*, 1997).

We have studied the effect of the *rac* mutation on the temporal and spatial expression patterns of three genes, *HRGPnt3*, *iaa4/5* and *gh3* previously shown to be expressed in adventitious root meristems (Hagen *et al.*, 1991; Ballas *et al.*,

1993; Vera *et al.*, 1994). This was done by comparing the temporal and/or spatial expression patterns of the three genes during early stages of cell division and root initiation in cuttings with wild-type versus *rac* background. Three genes were selected from previously isolated auxin-responsive or root-associated genes because they all have been shown to be expressed in adventitious root initials and transgenic tobacco lines with promoter–GUS fusions for each of the genes were available.

*iaa*4/5 and *gh*3 are thought to be early auxin-responsive genes because accumulation of each of their transcripts is rapid, specific for biologically active auxins and independent of protein synthesis. Transcriptional activation of both the *iaa*4/5 and *gh*3 promoters is detectable in a variety of auxin-responsive cell types in addition to adventitious and lateral root meristematic cells. Activation of *HRGPnt3* promoter sequences is only detectable in primary root meristematic cells or cells determined for secondary adventitious root formation. Since the *rac* mutation had been defined on a phenotypic but not on a molecular basis, the effects of *rac* on the transcriptional induction of *iaa*4/5, *gh*3 and *HRGPnt3* should provide information regarding *rac*'s role in auxin signal transduction for cell divisions and/or specifically for adventitious root initiation.

Using histochemical staining and quantitative analyses of *iaa*4/5, *gh*3-GUS and *gh*3-GUS transformant, wild-type and *rac* cuttings (Lund *et al.*, 1997), we determined that *rac* does not repress the auxin activation of the promoters of these two genes. This indicates that activation of the *iaa*4/5 and *gh*3 promoters is not limiting for adventitious root initiation in *rac* cuttings. Based on fluorometric analyses of *gh*3-GUS activity in *gh*3-GUS hemizygous cuttings versus *gh*3-GUS hemizygous, *rac* heterozygous cuttings treated with auxin for 1 day, we concluded that the *rac* mutation does not cause any reduction in IBA sensitivity at the level of *gh*3-GUS activity. Thus, these results do not support the conclusion of Barbier-Brygoo *et al.* (1990) that the *rac* mutation causes a general reduction in auxin sensitivity.

Using histochemical staining analyses of auxin-treated, *HRGPnt3*-GUS transformant, wild-type and *rac* cuttings (Lund *et al.*, 1997), we demonstrated that the *rac* mutation blocks activation of the 1.3-kb, 5' upstream region of the *HRGPnt3* promoter. Thus, activation of a region within the *HRGPnt3* promoter occurs specifically during adventitious root initiation in tobacco cuttings. RNA blot analyses, however, suggest that *HRGPnt3*-GUS expression is regulated both developmentally and environmentally. Based on the *HRGPnt3*-GUS analyses, we concluded that differential expression in response to auxin treatment occurs during adventitious root initiation in the wild-type versus callus formation in *rac* cuttings.

If adventitious root initiation is regarded as an organized form of cell division and auxin-induced adventitious root initiation occurs via auxin reception followed by transduction of the auxin signal, our data place *iaa*4/5 and *gh*3 expression upstream from *rac*, and *HRGPnt3* expression downstream from *rac* in a simple linear model. Although early auxin-responsive gene expression is likely to be a component of adventitious root initiation, it is doubtful that expression of genes such as *iaa*4/5 and *gh*3 have specific roles in the organization of the cell divisions that are required for adventitious root initiation.

Furthermore, it is unlikely that expression of upstream, auxin signal transduction genes has a role in the determination of competence for adventitious root initiation in cuttings. Conversely, it is plausible that genes such as *HRGPnt3* that code for structural proteins might be upregulated prior to, or coincident with, the organization of cell divisions that occurs during adventitious root initiation, and thus be a factor in the determination of the competence for adventitious root initiation in tobacco cuttings. Since the *rac* mutation has a major phenotypic effect of blocking adventitious root initiation prior to the first organized cell division, it is likely that *rac* has an important regulatory role during the early stages of adventitious root initiation in the phloem parenchyma or inner cortical parenchyma cells in tobacco cuttings.

Conclusions

Monogenic mutants of *A. thaliana* that do not form adventitious roots in response to auxin may be most useful in analysing the basis of adventitious rooting competence as it is lack of competence for rooting that is important horticulturally. Of the mutants currently identified, *rml* 2 and *alf* 4-1 may be particularly useful.

The *rty* mutants of *A. thaliana* demonstrate that production of large numbers of adventitious roots can be the result of high endogenous IAA levels, but they must also have high competence for adventitious root initiation.

In the *rac* mutant of tobacco, which is incompetent for auxin-induced root initiation but not cell division, activation of the promoters of the early auxin-responsive genes is not limiting for adventitious root initiation based on histochemical GUS staining and quantitative analyses of appropriate genetic transformants.

Based on histochemical staining of appropriate genetic transformants, we also conclude that the *rac* mutation blocks activation of the *HRGPnt3* promoter and that there is differential expression of the promoter in response to auxin treatment during adventitious root initiation in wild-type versus callus formation in *rac* cuttings.

Future Directions and Limitations

As indicated in the Introduction, a series of single gene mutants with developmental blocks in various steps in the root initiation process is desirable because rooting is likely to be a multistep process (Malamy and Benfey, 1997). Detecting mutants for the adventitious rooting process is a laborious and time-consuming undertaking even in a species like *A. thaliana*. What would make screening for mutants much easier is the insertion of a genetic modification into the population prior to mutagenesis treatment that would allow controlled lethality of all individuals except those that have a mutation in a known root initiation-related gene promoter. This, of course, would make the individuals with a root initiation-related mutation easily identifiable. The search for a method of controlled lethality is very important in order to enhance progress in understanding competence for adventitious rooting using mutants.

References

Ballas, N., Wong, L.M. and Theologis, A. (1993) Identification of the auxin-response element Aux RE in the primary indoleacetic acid-inducible gene, PS-Iaa 4/5, of pea (*Pisum sativum*). *Journal of Molecular Biology* 223, 580-596.

Barbier-Brygoo, H., Maurel, C., Shen, W.H., Ephritikhine, H., Delbarre, A. and Guern, C. (1990) Use of mutants and transformed plants to study the action of auxins. In: Hooley, R. and Roberts, J. (eds), *Hormone Perception and Signal Transduction in Animals and Plants*. Society of Experimental Biology Symposium, Company of Biologists, Cambridge, pp. 67-77.

Bentley, R. (1990) The shikimate pathway - a metabolic tree with many branches. *Critical Reviews in Biochemical Molecular Biology* 25, 307-384.

Boerjan, W., Cervena, M.J., Delarue, M., Beeckman, T., Dewittes, W., Bellini, C., Caboche, M., Van Onckelen, H., van Montague, M. and Inze, D. (1995) *Superroot*, a recessive mutation in *Arabidopsis*, confers auxin overproduction. *The Plant Cell* 7, 1405-1419.

Caboche, M., Muller, J.F., Chanut, F., Aranda, G. and Cirakoglu, S. (1987) Comparison of the growth promoting activities and toxicities of various auxin analogs on cells derived from wildtype and a non-rooting mutant of tobacco. *Plant Physiology* 83, 795-800.

Celenza, J.L., Grisafi, P.L. and Fink, G.R. (1995) A pathway for lateral root formation in *Arabidopsis thaliana*. *Genes and Development* 9, 2131-2142.

Cheng, J.C., Seeley, K.A. and Sung, Z.R. (1995) *rml*.1 and *rml*.2, *Arabidopsis* genes required for cell proliferation at the root tip. *Plant Physiology* 107, 365-376.

Ernst, S.G. (1994) Model systems for studying adventitious root initiation. In: Davis, T.D. and Haissig, B.E. (eds), *Biology of Shoot-borne Root Formation*. Plenum Press, New York, pp. 77-86.

Hagen, G., Martin, G., Li, Y. and Guilfoyle, T.J. (1991) Auxin-induced expression of the soybean *gh*3 promoter in transgenic tobacco plants. *Plant Molecular Biology* 17, 567-579.

King, J.J., Stimart, D.P., Fischer, R.H. and Bleecker, A.B. (1995) A mutation altering auxin homeostasis and plant morphology in *Arabidopsis*. *The Plant Cell* 7, 2023-2037.

Lund, S.T., Smith, A.G. and Hackett, W.P. (1996) Cuttings of a tobacco mutant, *rac*, undergo cell division but do not initiate adventitious roots in response to exogenous auxin. *Physiologia Plantarum* 97, 372-380.

Lund, S.T., Smith, A.G. and Hackett, W.P. (1997) Differential gene expression in response to auxin treatment in the wildtype and *rac*, an adventitious rooting incompetent mutant of tobacco. *Plant Physiology* 114, 1197-1206.

Malamy, J.E. and Benfey, P.N. (1997) Organization and cell differentiation in lateral roots of *Arabidopsis thaliana*. *Development* 124, 33-44

Muller, J.F., Goujaud, J. and Caboche, M. (1985) Isolation *in vitro* of naphthalene acetic acid-tolerant mutants of *Nicotiana tabacum* which are impaired in root morphogenesis. *Molecular and General Genetics* 199, 194-200.

Vera, P., Lamb, P.J. and Doerner, P.W. (1994) Cell-cycle regulation of hydroxyproline-rich glycoprotein *HRGPnt3* gene expression during initiation of lateral root meristems. *The Plant Journal* 6, 717-727.

9 Physiological Analysis of the Floral Transition

G. Bernier, L. Corbesier, C. Périlleux, A. Havelange and P. Lejeune

Laboratoire de Physiologie Végétale, Université de Liège, Sart Tilman, B4000 Liège, Belgium

Introduction

In flowering physiology, the major enigma since the 1930s remains the elusive nature of the endogenous signal or signals which are produced in the leaves exposed to inductive photoperiodic conditions and then transported to the shoot apical meristem (SAM) where this or these signals cause the transition from leaf to flower morphogenesis.

An early theory proposed that the signal is simple, specific and universal (see Bernier *et al.*, 1981). This putative signal was believed to be a hormone and was named 'florigen'.

During the last 50 years, numerous efforts have been made to isolate this hormone either from induced leaves or induced apices. Also, since it was demonstrated that the floral signal(s) are transported in the phloem (Bernier *et al.*, 1981), another approach was to extract materials derived from the phloem sap (Cleland, 1978). As is well known, all these attempts have so far been unsuccessful (Bernier, 1988), and as a result most workers in the field have discontinued this kind of work.

In fact, the often unconfessed goal of these studies was to isolate a 'magic' molecule that would be capable of causing flower formation when applied alone to vegetative individuals of all plant species. So far, that kind of molecule does not seem to exist. Gibberellins which are closest to the definition of florigen are indeed capable of causing flower formation in a number of species but inhibit flowering or are without effect in other species (Bernier, 1988). Since movement of floral signal(s) is in the phloem, there have been surprisingly, almost no attempts in the past to analyse in detail the changes in contents of the phloem sap at the time of floral induction. Such an analysis is the topic of the present chapter. For the chemicals whose levels are found to change in the sap at floral induction, we shall determine in which part of the plant they originate before entering the leaf phloem.

Plant Materials

We have used several photoperiodic model species that can be induced to flower by a single favourable 24-h cycle: the long-day (LD) plants *Sinapis alba*, *Arabidopsis thaliana* 'Columbia' and *Lolium temulentum* 'Ceres' that can all be induced by exposure to a single LD, and the short-day (SD) plant *Xanthium strumarium* that can be induced by exposure to a single long night (LN). Interestingly, all of the above species that are induced by a single LD can also be induced by a single displaced short day (DSD), i.e. an SD given at an unusual time within a 24-h cycle. A DSD avoids the lengthening of the light period and this eliminates the complications related to the extended period of photosynthetic activity characteristic of the LD. All the plant materials were grown from sowing to the end of the experiment in the strictly controlled conditions of our phytotronic growth rooms.

Phloem Sap Collection

In all plant species, phloem exudate (sap) was collected from excised mature leaves using adaptations of the ethylene diamine tetraacetic acid (EDTA) method of King and Zeevaart (1974). Adaptations were slightly different for the different species (see the details in the papers quoted below).

The phloem sap is a complex mixture of various classes of chemicals. Only five of these classes have been studied so far: carbohydrates, cytokinins, amino acids, polyamines and inorganic ions.

Carbohydrates

The carbohydrates of the sap were analysed using high-performance liquid chromatography (HPLC) and refractometry. In all investigated species, sucrose is the major carbohydrate of the phloem sap. Our results indicate that in all the above species the level of sucrose increases temporarily and markedly in the sap of plants induced to flower compared with non-induced plants (see Lejeune *et al.* (1991, 1993) for *S. alba*; Corbesier *et al.* (1998) for *A. thaliana*; Périlleux and Bernier (1997), for *L. temulentum*; Houssa *et al.* (1991) for *X. strumarium*). Remarkably, these increases are observed not only in inductive conditions involving a lengthening of the period of photosynthetic CO_2 fixation (exposure to a LD) but also in conditions where induction does not modify (exposure to a DSD) or decreases photosynthetic CO_2 fixation (exposure to an LN).

These increases in sucrose level in the sap occur very early in all cases and, except in *L. temulentum* induced by an LD, at times quite compatible with the timing of export of the floral signal(s) out of induced leaves, as determined from sequential defoliation experiments (see Bernier (1989) for *S. alba*; Corbesier *et al.* (1996) for *A. thaliana*; Périlleux *et al.* (1994) for *L. temulentum*; Houssa *et al.* (1991) for *X. strumarium*). The possible reason for the apparently contradictory behaviour of *L. temulentum* when induced by an LD (not when induced by a DSD) is discussed by Périlleux and Bernier (1997).

The problem of whether the extra sucrose of the sap at floral induction comes from recent photosynthesis or from reserve (starch) mobilization has been discussed for many years. Use of the *pgm* mutant of *A. thaliana*, which is unable to synthesize starch and thus possesses only 5–10% of the starch level found in the wild-type plants, has produced an answer to this question (Corbesier *et al.*, 1998). The *pgm* plants exposed to an LD flower almost as well as the wild-type plants and exhibit a similar increase in the sucrose level in the phloem sap. On the contrary, *pgm* plants exposed to a DSD flower weakly compared with wild-type plants and do not exhibit the increase in sap sucrose characteristic of wild-type plants. From these observations, we conclude that when induction involves an extended period of photosynthesis (exposure to an LD), the extra sucrose comes from recent photosynthesis, whereas when the period of photosynthesis is unaffected at induction (exposure to a DSD), the extra sucrose comes from starch mobilization. Starch mobilization is also presumably involved in the case of SD plants, such as *X. strumarium*, where induction is by a reduction in the period of photosynthesis.

In *S. alba* plants induced by a DSD, when the sucrose levels are monitored comparatively in the sap exported by mature leaves and in the sap reaching the SAM, it appears that starch mobilization occurs not only in leaves but also in other parts of the plant, either the stem or the roots, or both (Lejeune *et al.*, 1993).

Cytokinins

Cytokinins present in the sap were analysed using HPLC and radioimmunoassay (Lejeune *et al.*, 1994). The major cytokinin in the phloem sap of *S. alba* was either isopentenyladenine or its riboside, depending on the experiment. As was the case for sucrose, floral induction by an LD causes an increase in cytokinin levels in the sap, but this increase seems to occur a few hours later than the rise of sucrose (Bernier *et al.*, 1990). Similar observations have been made in *A. thaliana* 'Columbia' induced to flower by an LD (L. Corbesier, unpublished). In *X. strumarium* induced by an LN, the cytokinin content of the sap first decreases in the hours just following the LN and then increases moderately and transiently (Kinet *et al.*, 1994). As in *S. alba*, the increased cytokinin export from leaves occurs later than that of sucrose, but it still takes place at a time corresponding to the export of the floral stimulus out of induced leaves (Houssa *et al.*, 1991).

The extra cytokinins exported by induced *S. alba* leaves seem to originate in roots, since we detected an early stimulation of cytokinin export (in the xylem sap) by roots following floral induction (Bernier *et al.*, 1990; Lejeune *et al.*, 1994). Bark ringing between the mature leaves and the root system indicates that this stimulation of cytokinin export is caused by a foliar signal transported very early in the phloem sap to the roots (Bernier *et al.*, 1993). The bark ringing experiments also demonstrate that the leaf to root movement of this signal is essential for flowering. Further experiments have demonstrated that this signal is nothing other than the extra sucrose shown above to be released by induced leaves (A. Havelange et al., unpublished).

The extra cytokinins found in the phloem sap exported by *X. strumarium* leaves are certainly not of root origin since the cytokinin levels in the xylem sap exported by roots decrease markedly at induction in this species (Wareing et al., 1977; Kinet et al., 1994). In this case, the extra cytokinins apparently originate in the leaves themselves, perhaps by mobilization of storage forms of these compounds.

Amino Acids and C:N Ratio

Analysis of amino acids of the phloem sap was by HPLC and fluorimetry after 9-fluorenylmethoxycarbonyl chloride (FMOC) derivatization. In both *S. alba* and *A. thaliana*, there are five major amino acids in the sap: glutamine, glutamic acid, asparagine, aspartic acid and serine. In both species, induced to flower by either an LD or a DSD, Corbesier et al. (unpublished) have found marked increases in the total amount of amino acids present in the sap. In *S. alba*, the increased export of amino acids starts some hours after the increased export of carbohydrates, whereas in *A. thaliana* it occurs at about the same time. In *S. alba* at least, the increased amino acids are of leaf origin since their levels, as well as the nitrate level, were found to decrease in the xylem sap moving from roots to leaves at floral induction.

The carbohydrate:amino acid ratio (or C:N ratio) in the sap exported by leaves was found to increase two- to fourfold and at early times as a result of floral induction in both *S. alba* and *A. thaliana*. Thus, the supplies of both C and N to the shoot apex are increased at induction but the C supply is more stimulated than the N supply.

Polyamines

As shown by thin-layer chromatography (TLC) and fluorimetric analysis, the only polyamine present (as free and conjugated forms) in the sap of *S. alba* is putrescine (Havelange et al., 1996). As for amino acids, the levels of the two forms of putrescine increase early and markedly in the sap of *S. alba* plants induced to flower by an LD or a DSD. Since: (i) the putrescine level in the xylem sap moving from roots to leaves is low, and (ii) there is no increase in the putrescine level of this sap at induction, the extra putrescine is synthesized in the induced leaves themselves, just like the extra amino acids. The increased export of putrescine from induced leaves occurs at about the same time as the increased export of cytokinins.

Inorganic Ions

Not all components of the phloem sap change their levels at floral induction. Indeed, in the sap of *S. alba* plants induced by either an LD or a DSD, there are no detectable changes in the levels of three cations (Ca^{2+}, Mg^{2+}, K^+) and one anion (NO_3^-) (Havelange and Bernier, 1993; L. Corbesier, unpublished).

Conclusions

With regard to the (still incomplete) experimental evidence, it appears that:

1. Floral induction in all plant species causes a multiplicity of important, sometimes dramatic, changes in the contents of the phloem sap exported by leaves. These changes affect almost all kinds of chemicals analysed, including carbohydrates, cytokinins, amino acids and polyamines. There is no doubt that other chemical classes which change remain to be identified.

2. All these changes in phloem sap contents occur sufficiently early (except for the sucrose increase in *L. temulentum* induced by an LD) to be compatible with the timing of export of the floral stimulus out of induced leaves.

These physiological observations alone are insufficient, of course, to conclude that all the changing chemicals of the phloem sap are real floral signals. They simply suggest that these chemicals are potential floral signals and need to be substantiated by genetic studies. Although the genetic study of flowering time control, using well-defined mutant and transgenic plants, is only beginning, some results already support the suggestions arising from the above physiological work. For example, transgenic plants in which the sucrose availability is increased flower earlier than their wild-types, as is the case for potato plants in which starch biosynthesis is inhibited in leaves and roots due to suppression of ADP-glucose pyrophosphorylase (Müller-Röber *et al.*, 1992), and tomato plants in which sucrose phosphate synthase is overexpressed (Micallef *et al.*, 1995). Similarly, transgenic plants of *Lotus corniculatus* overexpressing in their shoot a soybean gene encoding a cytosolic glutamine synthetase and thus enriched in amino acids flower earlier than untransformed plants (Vincent *et al.*, 1997). These results clearly support the idea that sucrose and amino acids are real floral signals or, in other words, are components of the floral stimulus, at least in some plants.

3. In all cases, the sucrose increase clearly precedes the start of SAM activation, as evidenced by the persisting stimulation of cell divisions (Bernier *et al.* (1967) for *S. alba*; A. Jacqmard (unpublished data) for *A. thaliana*; Ormrod and Bernier (1990) for *L. temulentum*; Jacqmard *et al.* (1976) for *X. strumarium*). Thus, the export of more sucrose by leaves is apparently not the result of a higher demand (sink activity) of the SAM. This suggests that the extra sucrose plays a signalling role.

4. In *S. alba* and *X. strumarium*, at least, the increased export of sucrose by leaves precedes that of other potential floral signals, such as cytokinins or amino acids. The extra sucrose in *S. alba* was found to be the signal causing an increased flux of cytokinins from roots to shoot. Thus, at least in certain cases, there is clearly a sequence in the supply of floral signals from leaves to SAM.

5. Different floral signals originate in different parts of the plant. If sucrose, amino acids and putrescine are clearly of leaf origin in *S. alba*, cytokinins, on the contrary, are most likely of root origin. The parallel increases in amino acid and putrescine export from induced leaves of *S. alba* plants can presumably be explained by the fact that putrescine is known to derive from glutamic acid.

6. Taken as a whole, the results presented here support the theory of the multifactorial control of floral transition (Bernier, 1988).

Acknowledgements

This research was supported by grants from the 'Pôles d'Attraction Interuniversitaires' pour le Compte de l'Etat Belge, Services du Premier Ministre - Services Fédéraux des Affaires Scientifiques, Techniques et Culturelles (P 4/15) and an EC Project no. BIO4CT97-2231. L.C. is grateful to the F.R.I.A. for the award of a research fellowship.

References

Bernier, G. (1988) The control of floral evocation and morphogenesis. *Annual Review of Plant Physiology and Plant Molecular Biology* 39, 175-219.

Bernier, G. (1989) Events of the floral transition of meristems. In: Lord, E.M. and Bernier, G. (eds), *Plant Reproduction: From Floral Induction to Pollination*. The American Society of Plant Physiologists Symposium, Vol. 1, Rockville, Maryland, pp. 42-50.

Bernier, G., Kinet, J.-M. and Bronchart, R. (1967) Cellular events at the meristem during floral induction in *Sinapis alba* L. *Physiologie Végétale* 5, 311-324.

Bernier, G., Kinet, J.-M. and Sachs, R.M. (1981) *The Physiology of Flowering*, Vol. I. CRC Press, Boca Raton, Florida.

Bernier, G., Lejeune, P., Jacqmard, A. and Kinet, J.-M. (1990) Cytokinins in flower initiation. In: Pharis, R.P. and Rood, S. (eds), *Plant Growth Substances 1988*. Springer-Verlag, Berlin, pp. 486-491.

Bernier, G., Havelange, A., Houssa, C., Petitjean, A. and Lejeune, P. (1993) Physiological signals that induce flowering. *The Plant Cell* 5, 1147-1155.

Cleland, C.F. (1978) The flowering enigma. *BioScience* 28, 265-269.

Corbesier, L., Gadisseur, I., Silvestre, G., Jacqmard, A. and Bernier, G. (1996) Design in *Arabidopsis thaliana* of a synchronous system of floral induction by one long day. *The Plant Journal* 9, 947-952.

Corbesier, L., Lejeune, P. and Bernier, G. (1998) The role of carbohydrates in the induction of flowering in *Arabidopsis thaliana*: comparison between the wild type and a starch-less mutant. *Planta* 206, 131-137.

Havelange, A. and Bernier, G. (1993) Cation fluxes in the saps of *Sinapis alba* during the floral transition. *Physiologia Plantarum* 87, 353-358.

Havelange, A., Lejeune, P., Bernier, G., Kaur-Sawhney, R. and Galston, A.W. (1996) Putrescine export from leaves in relation to floral transition in *Sinapis alba*. *Physiologia Plantarum* 96, 59-65.

Houssa, P., Bernier, G. and Kinet, J.-M. (1991) Qualitative and quantitative analysis of carbohydrates in leaf exudate of the short-day plant, *Xanthium strumarium* L. during floral transition. *Journal of Plant Physiology* 138, 24-28.

Jacqmard, A., Raju, M.V.S., Kinet, J.-M. and Bernier, G. (1976) The early action of the floral stimulus on mitotic activity and DNA synthesis in the apical meristem of *Xanthium strumarium*. *American Journal of Botany* 63, 166-174.

Kinet, J.-M., Houssa, P., Requier, M.-C. and Bernier, G. (1994) Alteration of cytokinin levels in root and leaf exudates of the short-day plant *Xanthium strumarium* in response to a single long night inducing flowering. *Plant Physiology and Biochemistry* 32, 379-383.

King, R.W. and Zeevaart, J.A.D. (1974) Enhancement of phloem exudation from cut petioles by chelating agents. *Plant Physiology* 53, 96-103.

Lejeune, P., Bernier, G. and Kinet, J.-M. (1991) Sucrose levels in leaf exudate as a function of floral induction in the long day plant *Sinapis alba*. *Plant Physiology and Biochemistry* 29, 153-157.

Lejeune, P., Bernier, G., Requier, M.-C. and Kinet, J.-M. (1993) Sucrose increase during floral induction in the phloem sap collected at the apical part of the shoot of the long-day plant *Sinapis alba*. *Planta* 190, 71-74.

Lejeune, P., Bernier, G., Requier, M.-C. and Kinet, J.-M. (1994) Cytokinins in phloem and xylem saps of *Sinapis alba* during floral induction. *Physiologia Plantarum* 90, 522-528.

Micallef, B.J., Haskins, K.A., Vanderveer, P.J., Roh, K.S., Shewmaker, C.K. and Sharkey, T.D. (1995) Altered photosynthesis, flowering, and fruiting in transgenic tomato plants that have an increased capacity for sucrose synthesis. *Planta* 196, 327-334.

Müller-Röber, B., Sonnewald, U. and Willmitzer, L. (1992) Inhibition of the ADP-glucose pyrophosphorylase in transgenic potatoes leads to sugar-storing tubers and influences tuber formation and expression of tuber storage protein genes. *EMBO Journal* 11, 1291-1238.

Ormrod, J.C. and Bernier, G. (1990) Cell cycle patterns in the shoot apex of *Lolium temulentum* L. cv. Ceres during the transition to flowering following a single long day. *Journal of Experimental Botany* 41, 211-216.

Périlleux, C. and Bernier, G. (1997) Leaf carbohydrate status in *Lolium temulentum* L. during the induction of flowering. *New Phytologist* 135, 59-66.

Périlleux, C., Bernier, G. and Kinet, J.-M. (1994) Circadian rhythms and the induction of flowering in the long-day grass *Lolium temulentum* L. *Plant, Cell and Environment* 17, 755-761.

Vincent, R., Fraisier, V., Chaillou, S., Limani, M.A., Deleens, E., Phillipson, B., Douat, C., Boutin, J.-P. and Hirel, B. (1997) Overexpression of a soybean gene encoding cytosolic glutamine synthetase in shoots of transgenic *Lotus corniculatus* L. plants triggers changes in ammonium assimilation and plant development. *Planta* 201, 424-433.

Wareing, P.F., Horgan, R., Henson, I.E. and Davis, W. (1977) Cytokinin relations in the whole plant. In: Pilet, P.E. (ed.), *Plant Growth Regulation*. Springer-Verlag, Berlin, pp. 147-153.

10 Genetic and Environmental Control of Flowering in Strawberry

N.H. Battey[1], P. Le Mière[1], A. Tehranifar[1], C. Cekic[1], S. Taylor[1], K.J. Shrives[1,*], P. Hadley[1], A.J. Greenland[2], J. Darby[3] AND M.J. Wilkinson[1]

[1]*School of Plant Sciences, The University of Reading, Whiteknights, Reading RG6 6AS, UK;* [2]*ZENECA Agrochemicals, Jealott's Hill Research Station, Bracknell RG42 6EY, UK; and* [3]*Darby Brothers Farms Ltd, Methwold Hythe, Norfolk IP26 4PU, UK*

Introduction

There are two especially good reasons for understanding flowering in strawberry. First, flowering time and pattern determine, to a large extent, the time, duration and intensity of fruiting. Current fruit production methods, particularly for out-of-season crops, therefore revolve around control of the flowering process. Conversely, strawberry plant producers (propagators) need to control flowering in order to optimize runner production, because flowering and runnering are complementary aspects of development in the strawberry plant.

This leads to the second reason for understanding flowering in strawberry: not only is the plant very sensitive to the environment at all phases of its yearly cycle, and therefore a paradigm for environmental control of developmental physiology, but vegetative and floral development are also demonstrably under separate but interlinked control. This linkage is frequently (though not invariably) found in trees, for instance in apple (Mika, 1969, 1986) and larch (Longman *et al.*, 1965), and suggests that antagonism between vegetative and floral development may be of general importance in perennials. The strawberry is therefore a good model for both commercially and scientifically orientated studies of perennial flowering.

The aim here is to summarize our work on the mechanisms underlying the environmental and genetic control of flowering in strawberry. The emphasis of our applied research is to optimize control of flowering and cropping by quantifying environmental effects on growth and development. In our more fundamental research, the relatively simple genetic control of flowering in the diploid wild strawberry *Fragaria vesca* is being used to improve our under-

* Present address: Farms Advisory Services Team (FAST), Faversham, Kent ME13 0LN, UK.

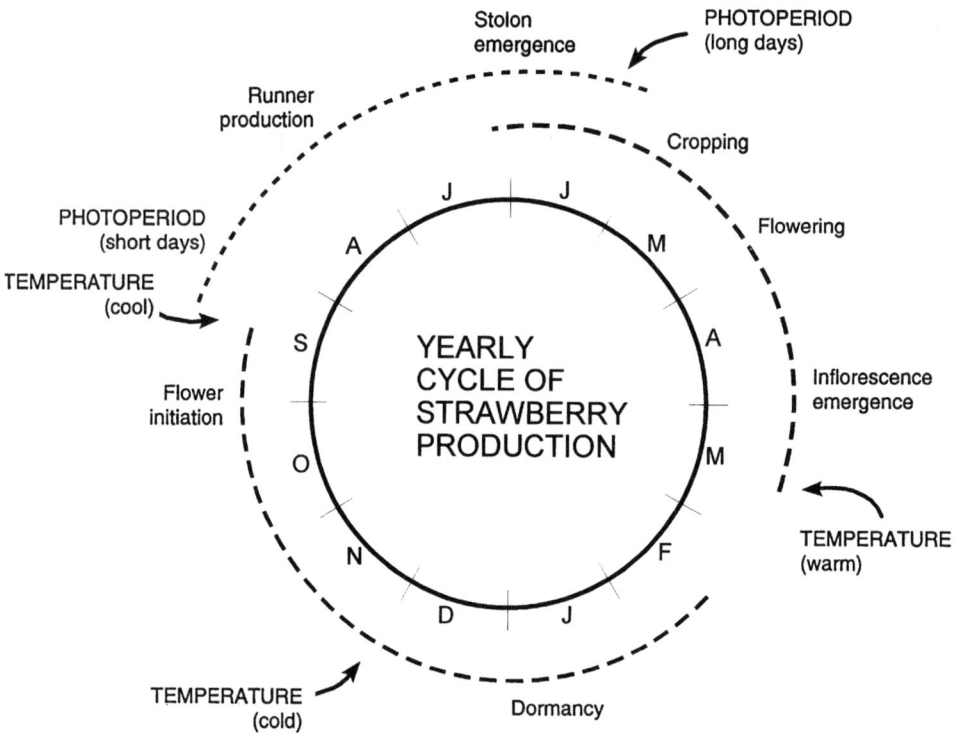

Fig. 10.1. The perennial cycle dictating fruit production and runner propagation in the June-bearing strawberry. The months of the year are indicated by their initial letters, and read anticlockwise.

standing of perennial flowering at the molecular level. This understanding has the potential to allow control of flowering and runnering in commercial strawberry without the current dependence on environmental cues.

Environmental Control of Growth and Development in Strawberry

The focus of our work is the main season or June-bearing strawberry cultivars, in particular the cultivar Elsanta, and a generalized illustration of their perennial cycle is given in Fig. 10.1. Environmental signals (e.g. cold temperatures) operate through (usually speculative) physiological mechanisms (e.g. floral inhibitor production) to bring about morphological responses (e.g. promotion of vegetative growth).

Flower induction and initiation

Flower induction in June-bearing strawberries occurs in the late summer and is controlled by temperature and photoperiod, so that at above about 15°C short days are required for flower initiation, whereas below this temperature the

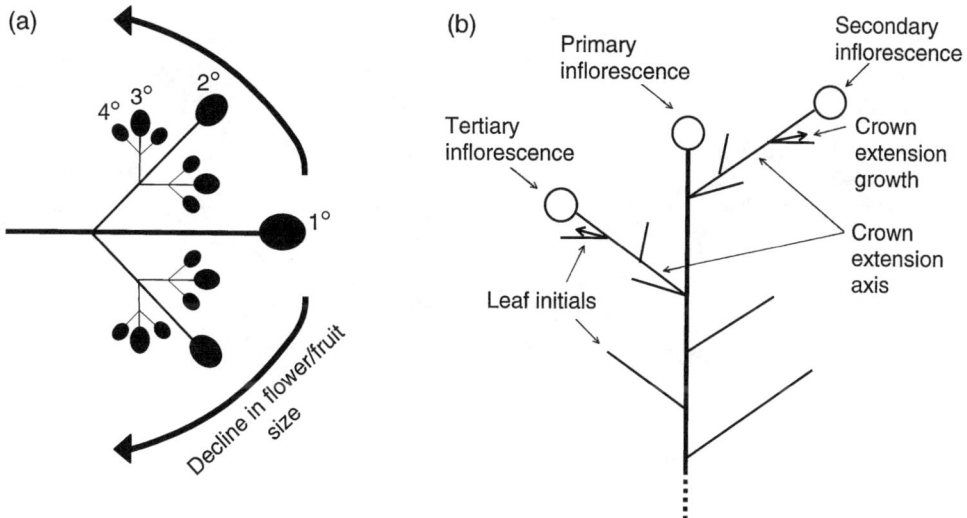

Fig. 10.2 (a) The sequence of flower initiation in the strawberry inflorescence. The positions of the primary, secondary, tertiary and quarternary flowers are indicated. (b) The sequence of inflorescence initiation in the strawberry crown.

plants flower without regard to photoperiod (Ito and Saito, 1962; Guttridge, 1985). There is considerable variation between cultivars in their response to photoperiod and temperature (Heide, 1977), but our work with cv. Elsanta indicates that in late summer about 3 weeks of inductive conditions are needed, and the first flower initials are clearly visible about 6 weeks later (J. Warnett, P. Le Mière and N.H. Battey, unpublished). The amount of induction required, and the time taken to initiate the flowers, are both strongly influenced by the time of year: immediately after chilling, plants are relatively insensitive to induction (see below).

Once the plants have been induced, flower initiation continues during the autumn, and our experiments indicate that temperature has the dominant role in controlling the rate, and therefore the extent, of flower initiation (Le Mière *et al.*, 1996). The strawberry inflorescence is an irregular dichasial cyme, which means that flowers are initiated in the sequence illustrated in Fig. 10.2a. Inflorescences (trusses) are initiated in basipetal sequence on successive branch crown extension axes (Fig. 10.2b). In cv. Elsanta there seems to be an upper limit of approximately 16 flowers in the inflorescence (Le Mière *et al.*, 1996). The inflorescences themselves, on the other hand, continue to form as long as growth continues, so that quaternary and quinary inflorescences are not uncommon.

In the field, the process of flower initiation is arrested by cold temperatures during the onset of winter. This arrest is an expected effect of cold temperature on growth rate. However, it is often loosely referred to as 'dormancy'. In contrast, our evidence suggests that there is a restraint on vegetative growth that accompanies the induction of flowering in September (Fig. 10.3). This is

18°C/15 h   12°C/9 h   18°C/15 h   Outside

(Control)   (After induction in early September)

Fig. 10.3. The restraint on growth that accompanies flower induction in cv. Elsanta. The control treatment was maintained in warm long days (18°C/15 h) from the beginning of September, delaying flower initiation by approximately 2 months. Note the greatly elongated petioles and upright growth habit. Plants in the other treatments were induced by shortening daylengths and cooler temperatures outside in early September. They were then placed into cool short days (12°C/9 h), warm long days (18°C/15 h), or left outside, until this photograph was taken (16 November). Note that the compact, flattened growth habit of the induced plants is maintained even if they are transferred into warm long days after induction.

measured by transferring plants to a favourable (warm) environment and recording their vegetative vigour; because some of this restraint is maintained whatever the external conditions, it represents 'true', 'innate' or 'endodormancy' (see Lang *et al.*, 1987, for terminology). Like flower initiation it is reversed by a prolonged period of cold or 'chilling'. This innate dormancy is only a partial restraint on growth in strawberry, but it and its antidote, chilling need to be understood in some detail because their effects are persistent and strongly influence flower emergence and cropping in the following season.

Chilling – promotion of vegetative growth, inhibition of flower initiation

Chilling in strawberry, as in many other perennials, is a key process. Its effect is to restore vegetative growth potential to its maximum and to stop further initiation of flowers. In the spring, when warm temperatures return, vegetative vigour is therefore very high, and existing flowers emerge.

The critical questions concerning chilling are:

- When do plants become sensitive to chilling?
- When is the chilling requirement saturated?
- What is the optimum chilling temperature; is the optimum the same for vegetative growth promotion, and subsequent flowering and fruiting?

Plants of the cv. Elsanta are certainly sensitive by November, and chilling sensitivity probably coincides with the acquisition of innate dormancy that accompanies flower induction in September (Fig. 10.4). The chilling requirement for vegetative vigour is saturated by 8 weeks at $-2°C$ in cold store, but the chilling requirement for optimum fruit set and cropping appears to be lower (4 weeks); this is probably because too much vegetative vigour depresses fruit set (Tehranifar *et al.*, 1998, and unpublished observations). Field chilling also seems to be much more effective than chilling given to plants in bags in the cold store (Tehranifar *et al.*, 1998, and unpublished observations). This may be because oscillating temperatures enhance the effectiveness of the chilling treatment; short periods of warm temperature are known to enhance chilling in peach trees (Erez and Couvillon, 1987). These general comments indicate the complexity of the response of cv. Elsanta to the cold. However, we now know enough to develop a provisional chill unit model that will allow prediction of dormancy breaking (chilling saturation) in cv. Elsanta, an important goal for optimizing out-of-season fruit production (see below). For other cultivars, only limited information is available, but this is enough to show large differences in chilling requirement (Bailey and Rossi, 1965; Kronenberg and Wassenaar, 1972; Avigdori-Avidov *et al.*, 1977; Craig and Brown, 1977).

Spring and summer – flower emergence, fruit growth and runner production

In the spring, the June-bearing strawberry undergoes active vegetative growth and flower initiation ceases. Cv. Elsanta appears to be insensitive to the inductive conditions (short days and temperatures below 15°C) that prevail in spring, for at least 4 weeks after chilling (A. Tehranifar, P. Le Mière and N.H. Battey, unpublished). This is consistent with data from other varieties, where the duration can be up to 9 weeks (Guttridge, 1958, 1985). It seems likely that by the time the plant becomes sensitive to induction, environmental conditions, particularly temperature, are usually no longer inductive. Instead, the warm temperatures and lengthening days promote runner production, which begins after flower emergence and fruit set. Runners develop from stolons, which are considered to be a specialized type of branch crown and are highly elongated at the first two internodes. The photo-thermal requirements for runner formation are cultivar specific, but generally long days greater than 14–16 h and temperatures above 17–20°C are optimal (Darrow, 1936; Smeets and Kronenberg, 1955; Heide, 1977).

Extensive experimentation under controlled environment conditions has revealed that in cv. Elsanta the time from the beginning of active vegetative growth to flowering and fruiting is strongly controlled by temperature (Le Mière *et al.*, 1998). Not only are the rates of progress to flowering and fruiting linearly related to temperature, but also the duration of fruiting and overall fruit yield decrease as temperature increases over the range 12–28°C (Fig. 10.5). Yield declines with increasing temperature apparently because canopy development is less at higher temperature, so that the plants have a smaller canopy at fruiting (Le Mière *et al.*, 1998). Hence, plants grown at higher temperatures can support a smaller crop than plants at lower temperatures. This quantitative information

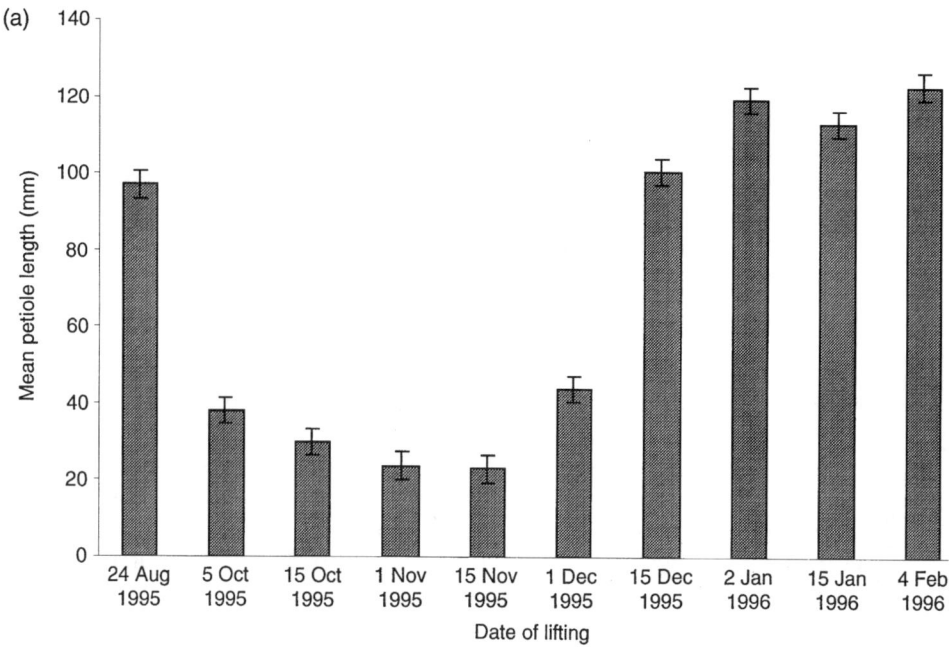

Fig. 10.4 (a) The development of a chilling requirement for maximum vegetative vigour in cv. Elsanta. Vegetative vigour can be assessed most easily by measuring petiole length (see Guttridge, 1985). Plants were lifted from the field at the intervals indicated, and grown for 40 days at 15°C/16 h daylength. The length of the petioles of the first three leaves to emerge was then recorded. Plants lifted between 5 October 1995 and 1 December 1995 have entered the dormant phase and require a period of chilling to attain maximum vegetative vigour. This requirement is fulfilled by 2 January 1996 in this experiment. (b) The appearance of plants lifted between 15 October 1995 and 2 January 1996 illustrates the effect of dormancy shown graphically in part (a).

Control of Flowering in Strawberry

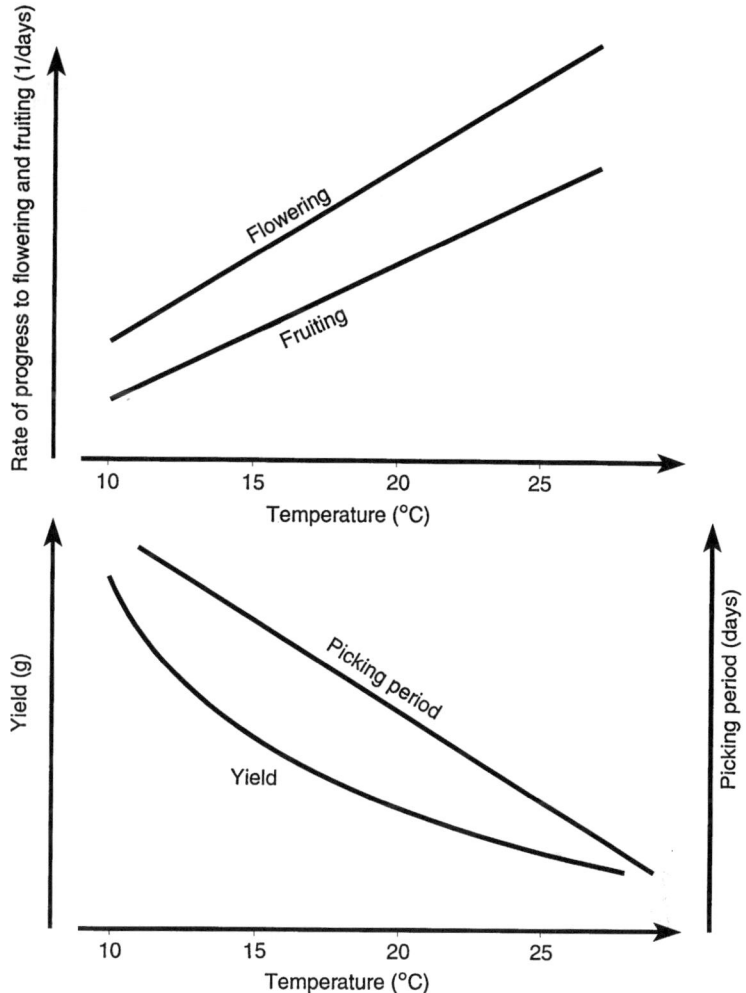

Fig. 10.5. The effect of temperature on the rate of approach to flowering and fruiting (upper), and the duration of fruiting and total yield (lower), in cv. Elsanta. For details see Le Mière et al. (1998).

provides the basis for predicting flowering and fruiting time, duration and intensity in cv. Elsanta; the effects of plant size and planting density on cropping pattern and yield can also be included.

The physiological basis of the perennial cycle in strawberry

Physiological studies in a range of species have shown that inductive conditions are perceived by the leaves and that the result is the production of a signal or signals; when transported to the meristem, this evokes flower initiation (Bernier et al., 1981; Thomas and Vince-Prue, 1996). Although in perennials the situation

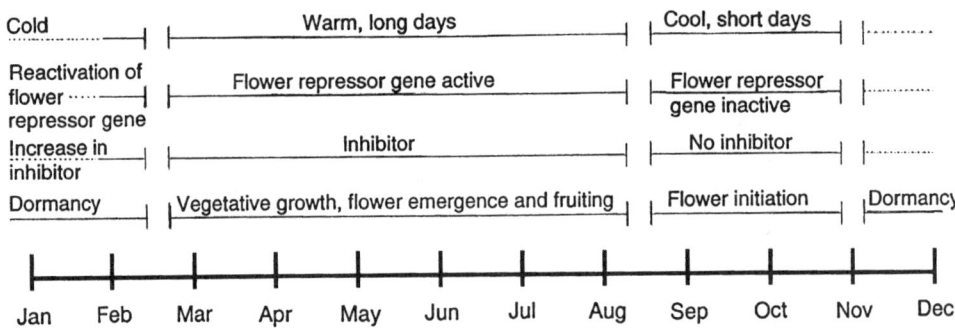

Fig. 10.6. The physiological and genetic control of the perennial cycle in the June-bearing strawberry.

is undoubtedly more complex than this (see, for example, Sedgeley, 1990), in the June-bearing strawberry circumstantial evidence suggests that the leaves produce an inhibitor of flowering in non-inductive conditions; flowering takes place when the leaves cease to produce the inhibitor as a result of induction (Guttridge, 1956, 1959a, b: reviewed in Guttridge, 1985). Detailed studies of the consequences of defoliation and other treatments suggest that promotive effects are at work as well as inhibitory ones (for review, see Durner and Poling, 1988). However, daylength extension experiments indicate that the pattern of sensitivity to red and far-red light in strawberry is similar to that found for promotion of flowering in long-day plants (Vince-Prue and Guttridge, 1973). This suggests that the (June-bearing) strawberry is a 'negative long-day plant', in which the environmental perception mechanism leads to production of an inhibitor, so preventing flowering in non-inductive long days. It is unclear exactly how the inductive control by temperature relates to that by photoperiod, but the same number of inductive cycles appears to be required for both signals, suggesting a common mechanism (see Heide, 1977). It is therefore possible to interpret the perennial cycle of flower initiation and vegetative growth in terms of cyclical inactivation/activation of inhibitor production (Fig. 10.6). Here the response of the plant to changing temperature plays a critical role in controlling inhibitor production. In the second part of this chapter, we describe our current efforts to define the molecular basis for this control.

The use of quantitative environmental response data to predict growth and cropping patterns in strawberry

The information presented in summary form above concerns the physiological responses of the strawberry to its environment. These basic responses have been well known for some time, worked out by Bringhurst, Guttridge, Poling and others (see reviews by Guttridge, 1985; Durner and Poling, 1988; Galletta and Bringhurst, 1990). The potential of the quantitative methodology we have employed is that it allows the development of mathematical models that can predict cropping patterns; this is of great value to commercial strawberry growers and propagators who need to programme production to meet market demand.

We have used the information on spring growth (Fig. 5; Le Mière *et al.*, 1998) to produce the predictive package BERRYSIM™. This uses the quantitative responses derived from our results to predict cropping profiles with increasing accuracy as environmental data from the current season are input by the user (Fig. 10.7). Our intention is to build on this basic package, so that ultimately it will include many aspects of crop management for different plant types, planting times and growing environments. For example, the chilling response of cv. Elsanta discussed above has important implications for cropping, particularly with out-of-season crops grown in polytunnels. Chilling duration has a direct effect on fruiting duration (Fig. 10.8), and we have a preliminary chill unit model (Fig. 10.9) that can be used to estimate the time to first fruiting (Fig. 10.10). In a similar way, the important effects of planting density on yield (Fig. 10.11) need eventually to be incorporated into BERRYSIM™. Of course, this predictive system is specific to cv. Elsanta. If this ceases to be an important cultivar, the models will need to be recalibrated for other promising June-bearing cultivars. The production of a similar framework for everbearing cultivars clearly needs to be investigated.

Genetic and Molecular Control of Flowering and Runnering in Strawberry

We have described our overall objectives as understanding perennial flowering, and using this understanding to control flowering and fruiting time, duration and intensity in strawberry. Here, our focus has been on a quantitative description of the key environmental effects on growth and development of the strawberry plant. In the longer term, one of our major goals is to allow control of flowering and runnering without the current dependence on environmental cues. Some of the objectives of existing breeding programmes are of this kind – for instance to introduce a continuous flowering habit to desirable June-bearing varieties such as cv. Elsanta, or, conversely, to introduce the free-running character into everbearers with good fruit characteristics. The problem has been that to introduce these characters without additional undesirable ones is very difficult, and it is here that a genetic/molecular approach is particularly powerful. Looking further ahead, the exogenous control of flowering and runnering, perhaps by application of chemicals which activate or repress the promoter controlling expression of the flowering and runnering genes, would be of considerable commercial importance. There are several examples of switchable promoters which could be used in this way in plants (Gatz, 1997). Here we outline our strategy for the isolation and cloning of the genes whose controlled expression ultimately could allow this specific control of flowering and runnering in strawberry.

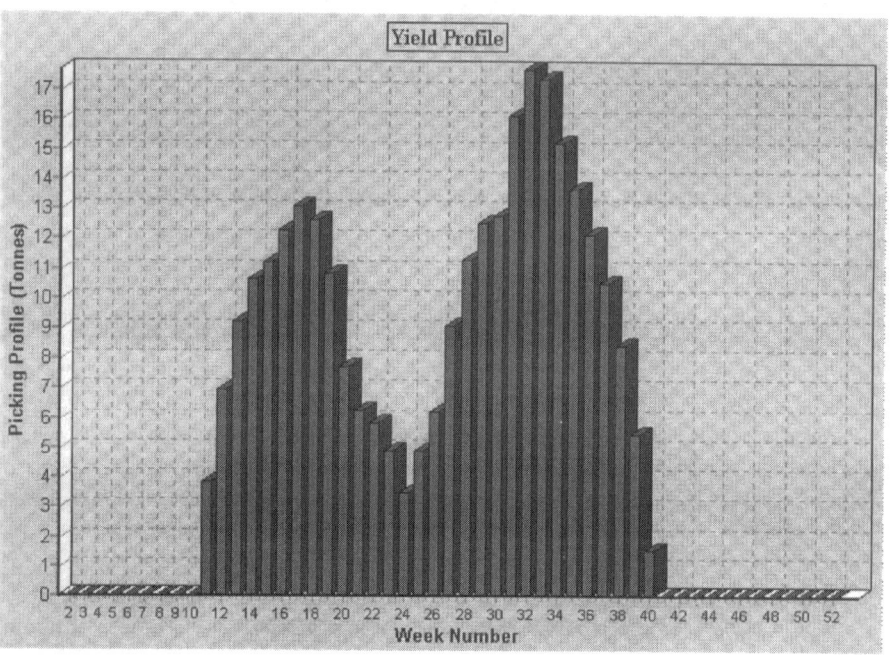

Fig. 10.7. BERRYSIM™, a computer package for predicting cropping times and patterns in cv. Elsanta. The upper panel shows the planting information input by the user and the calculated yield and picking date; the lower panel shows the predicted yield profile.

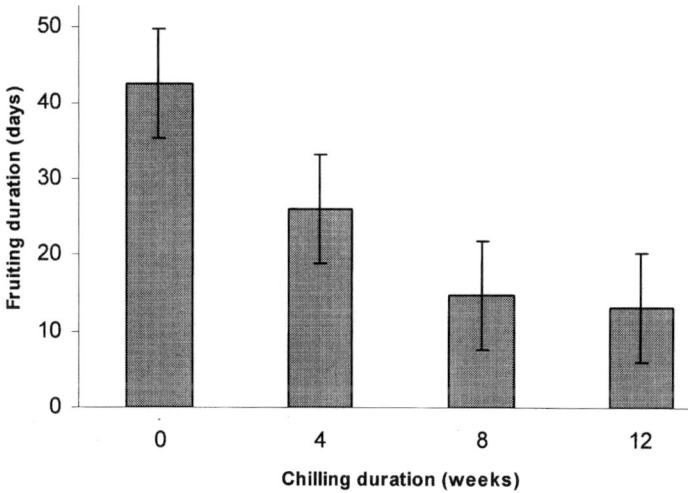

Fig. 10.8. The effect of chilling duration on fruiting duration in cv. Elsanta. After lifting in November and chilling for 0, 4, 8 or 12 weeks in a cold store at −2°C, waiting bed plants were potted and grown at 20°C/16 h daylength until fruiting. Bars indicate SED; d.f. = 8.

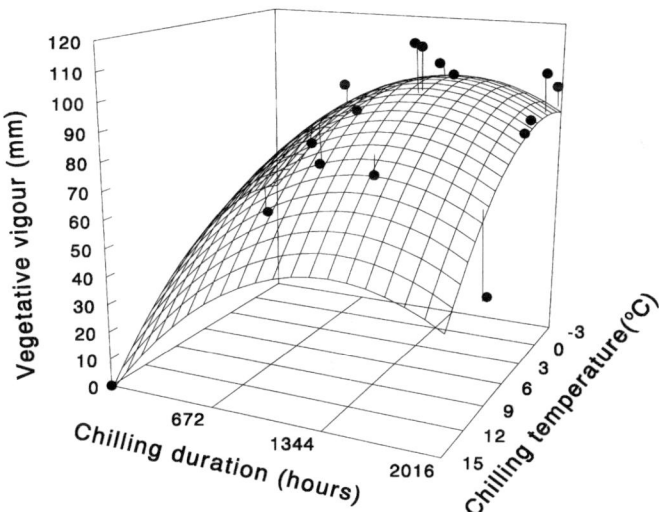

Fig. 10.9. A chill unit model for cv. Elsanta. The effect of chilling duration and chilling temperature on vegetative vigour (increase in mean petiole length) is shown. Waiting bed plants were lifted in November, chilled for 0, 4, 8 or 12 weeks at −2, −0.3, +3, +4 or +9.1°C, and then potted and grown at 15°C/16 h daylength for 90 days. At this time, the length of the petioles of the three oldest leaves unfolded after chilling were recorded. Vegetative vigour = $46.78 - 0.216\text{cht}^2 + 0.077\text{chd} - 0.000028\text{chd}^2$ ($r^2 = 0.96$). Cht, chilling temperature; chd, chilling duration. See also Tehranifar *et al.* (1998b).

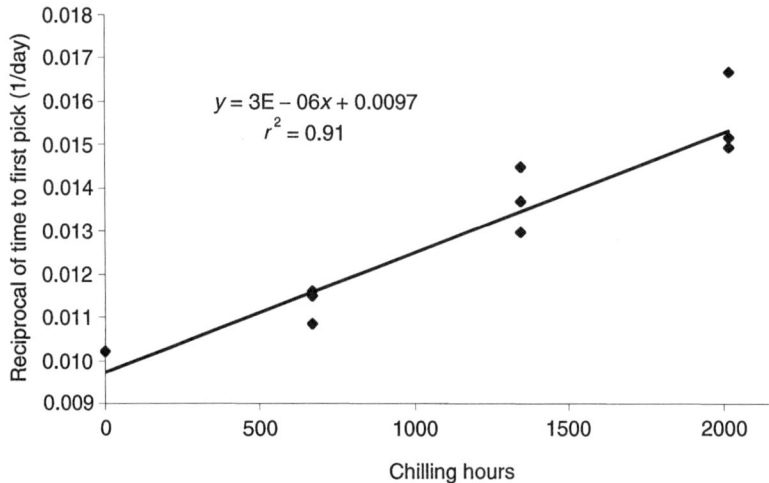

Fig. 10.10. Estimated time to fruiting based on chilling duration in cv. Elsanta. After lifting in November and chilling for 0, 4, 8 or 12 weeks in a cold store at −2°C, waiting bed plants were potted and grown until fruiting. The plants were grown under glass with supplementary lighting, to provide a daylength of 16 h.

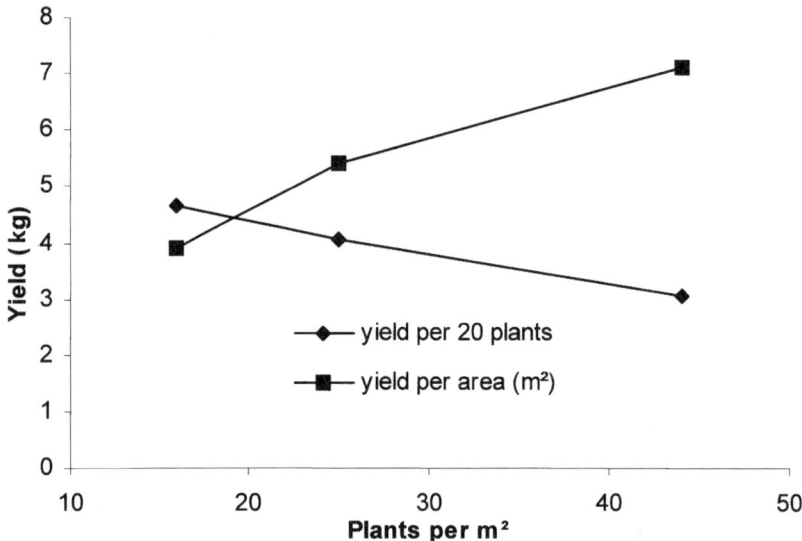

Fig. 10.11. The effects of planting density on yield in cv. Elsanta. Single-crowned waiting bed plants were planted in April at three densities and the total yield recorded. See also Tamiru (1995).

Genetic control of flowering and runnering

Genetic analysis of flowering in the commercial strawberry *F.* × *ananassa* is complex because it is an octoploid. It has been suggested that the genes for the everbearing trait have been introduced into octoploid cultivars from three independent sources (Sakin *et al.*, 1997); genetic segregation studies indicate that the everbearing (continuous flowering) character is dominant (Ourecky and Slate, 1967; Ahmadi *et al.*, 1990), and is conferred by one or more Mendelian genes. If this is correct, it would appear that the everbearing trait introduced into octoploid species evolved independently of that in diploid *F. vesca*, because here the situation is much more clear-cut. As with *F.* × *ananassa*, there are everbearing (*F. vesca semperflorens*) and seasonal flowering (*F. vesca* wild-type) forms of the species. *F. vesca semperflorens* produces no runners, whereas *F. vesca* is free-running. In a crucial paper, Brown and Wareing (1965) crossed *F. vesca* and *F. vesca semperflorens*, and showed that the dominant allele of a single gene confers seasonality to flowering; the dominant allele of an independently segregating gene allows runner production. Therefore the genetic evidence suggests that either different flowering pathways are affected in the octoploid and diploid strawberry, or genes controlling different steps in a single pathway are involved. On the other hand, the physiological basis of flowering in *F. vesca* and *F.* × *ananassa* is so similar that a common molecular basis is strongly implied. Only isolation and characterization of the genes controlling the perennial cycle of flowering will resolve this issue conclusively. The genetics of runner production have not been studied extensively; June-bearing cultivars are relatively free-running, whereas most everbearing cultivars produce only very few runners.

Our current objective is to isolate the dominant alleles controlling seasonal flowering and runnering in *F. vesca*, using a positional cloning approach. *F. vesca* is particularly well suited to this approach for the reasons indicated in Table 10.1. Dandekar's group in the USA has emphasized previously that these characteristics also make *F. vesca* an ideal model perennial plant, both for fundamental studies and for use in introgression work with octoploid strawberry cultivars (Uratsu *et al.*, 1991). There is enormous potential value in a perennial with many similarities in life cycle to large, long-lived and genetically

Table 10.1. *Fragaria vesca*: a model perennial plant.

Clear cycle of vegetative/reproductive growth
Genetics and physiology well-defined
Transformation system described
Diploid, $2n = 14$
Small genome size (2C estimated as ~ 0.8 pg)
Short generation time (~ 4 months)
Small, amenable to controlled environment work

complex tree species, but with a 4-month generation time (seed to seed and including scoring for flowering and runnering behaviour). Once the flowering and runnering genes have been isolated, our intention is to manipulate their expression levels to explore the possibility of molecular regulation of flowering pattern and runner production in *F. vesca*. A transformation system for *F. vesca* has been described (El Mansouri *et al.*, 1996). The possibility that these genes can be used to control flowering and runnering in *F.* × *ananassa* is clearly an exciting prospect, and will be investigated in parallel.

Fragaria vesca – genetics and molecular mapping

The results of our *F. vesca* × *F. vesca semperflorens* crossing programme confirm the findings of Brown and Wareing (1965). All the F_1 generation show seasonal flowering and are free-running, consistent with these being dominant characters (Fig. 10.12a). When the F_1 generation was back-crossed to *F. vesca semperflorens*, the BC_1 generation contained progeny in the expected ratio (1:1) for control of seasonal flowering and of runnering by independently segregating genes (Fig. 10.12b). We are now using a targeted approach for the positional cloning of these genes. High-resolution local maps around the two loci are needed, so that the most tightly linked markers can be used to screen a genomic library. Positive genomic clones will then be used to generate probes which, in turn, will be used to screen cDNA libraries. The practicality of this procedure rests on our ability to generate markers that are sufficiently closely linked to these two loci that they can be used to screen yeast artificial chromosome (YAC) colonies containing about 200 kb of genomic DNA or, preferably, DNA replacement vectors accommodating 9–23 kb of genomic DNA. This requires the selected markers to be positioned within approximately 15 kb of the target loci. The *F. vesca* genome has an estimated 2C value of 0.8 pg or a 1C value (representing the basic genome 'set') of 0.4 pg or 386 Mb (based on the closely related *Potentilla fruticosa*, Bennett and Leitch, 1995). It follows that 12,667 markers would have to be generated to provide a mean separation of 30 kb between neighbouring markers. This would ensure that any randomly selected locus within the genome would, in general, be no more than 15 kb from the nearest marker and, on average, would fall within 7.5 kb of the nearest marker (assuming a random distribution of markers).

We are using inter-simple sequence repeat-polymerase chain reaction (ISSR-PCR) (Charters *et al.*, 1996) to generate the molecular markers. Preliminary data led to the expectation that approximately 850 ISSR primers or pairs of ISSR primers (used in combination) would be required to generate the required number of markers. This number is too large to map directly, so candidate linked markers will be identified initially by bulk segregant analysis using a small subgroup of the main mapping population. Pooled DNA samples from 100 plants containing the target alleles and 100 plants lacking the alleles will be compared to identify markers within approximately 5 cM of the loci. The segregation of these markers will then be surveyed in detail in the 200 plant subpopulation. Only those markers present in all plants that contain each trait (flowering or runnering), and absent from all those lacking the trait, will be

Control of Flowering in Strawberry 125

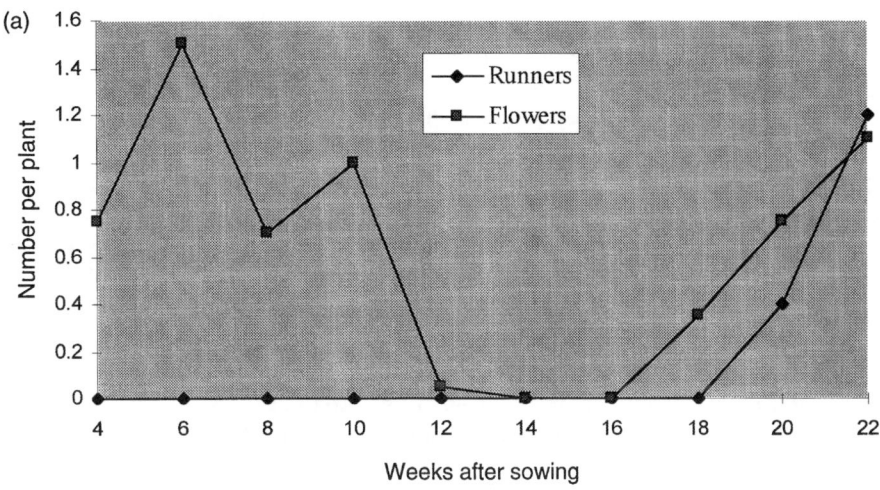

	Everbearing/ runnering	Everbearing/ non-runnering	Seasonal/ runnering	Seasonal/ non-runnering
BC₁ Expected %	25	25	25	25
Actual %	21	30	23	27
(Actual number	95	137	104	125)

Fig. 10.12 (a) F₁ progeny of *F. vesca* × *F. vesca semperflorens* all show seasonal flowering and are free-running. (b) Segregation of the runnering and seasonal flowering characters in the BC₁ generation. F₁ hybrids were back-crossed to the *F. vesca semperflorens* parent and yielded progeny that showed the expected segregation for dominance of seasonal flowering over everbearing and free-runnering over non-runnering.

selected. These markers will all lie within 0.5-1 cM of the target gene.

In the absence of either a physical or linkage map of the species, it is difficult to relate this linkage distance to physical separation. A broad estimate can be made, however, because physical distance and linkage are both roughly correlated to genome size. The *F. vesca* haploid genome (0.4 pg) falls between that of *Brassica rapa* (0.5 pg) and *Arabidopsis* (0.2 pg), in which 1 cM has been calculated to represent about 700 kb (Sadowski *et al.*, 1996) and 185 kb respectively (Schmidt *et al.*, 1995). In *F. vesca*, therefore, a reasonable estimate would be that 1 cM is roughly equivalent to 500 kb. Thus, a 0.5-1 cM resolution would represent approximately 250-500 kb. The segregation of selected markers will be studied on the mapping population of about 3000 plants from the first back-cross generation to produce a detailed 'local' map around the targeted loci and to identify markers most closely linked to the two traits. These will then be used to screen the genomic library and thence the cDNA libraries.

In order for this strategy to succeed, it is important that the chosen marker system reveals polymorphisms between the two parents and that the profiles of the F_1 hybrids are additive (i.e. they contain only bands deriving from both parents). Our preliminary screening using ISSR primers has shown that they do reveal polymorphisms between the two forms of *F. vesca* and that the bands from both parents are represented in the F_1 progeny (Fig. 10.13). The levels of homozygosity in the parental lines determine the simplicity of data analysis in the mapping population. The use of near isogenic lines as parents would generate a uniform (and highly heterozygous) F_1 and would also lead to segregation for nearly all polymorphic loci in the test BC_1 population. The homozygosities of the *F. vesca* and *F. vesca semperflorens* lines used as original

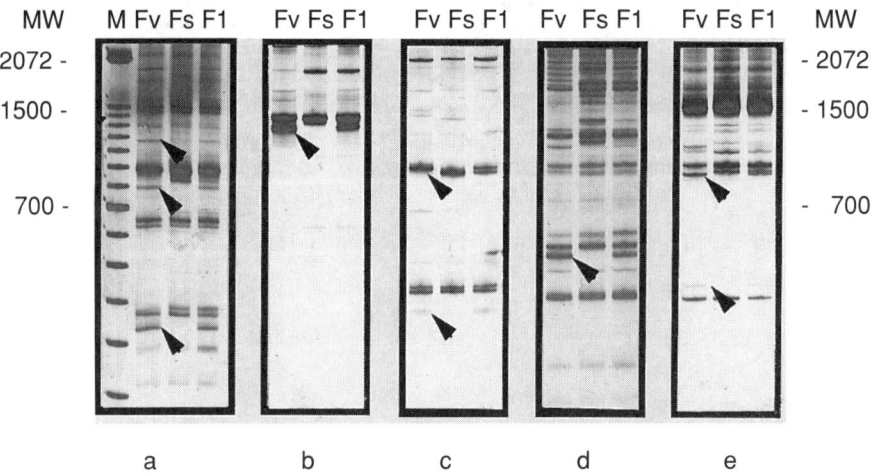

Fig. 10.13. ISSR-PCR analysis of *F. vesca* (Fv) and *F. vesca semperflorens* (Fs) reveals *F. vesca*-specific markers (arrowed) inherited by the F_1 hybrids (F1). The primers used were: (a) 807; (b) 835; (c) 844; (d) 888; (e) 841, from UBC set 9.

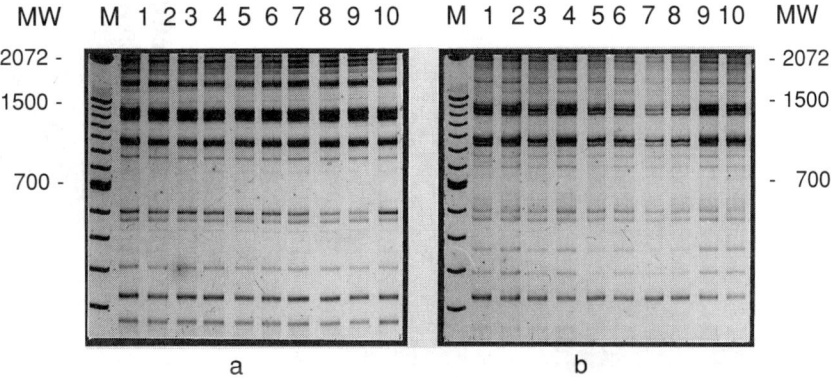

Fig. 10.14. Example ISSR-PCR analyses of self-pollinated progenies of the *F. vesca* (a) and *F. vesca semperflorens* (b) parental lines. The absence of segregation in either progeny is demonstrated in these gels by the consistency of banding profiles across all offspring. Primer 890 from UBC set 9 was used for these analyses.

parents in the current study were assessed using selfed progenies from each parent (30 progeny from each). The four ISSR primers used generated over 100 band positions in the parents and the progenies. None of these loci segregated amongst the progeny of either parent (Fig. 10.14) indicating that both parents are sufficiently homozygous to be considered near isogenic (>94% homozygous). This enables us to use the BC_1 generation as the mapping population.

Fragaria vesca – physiology

Control of flowering and runnering in *F. vesca* appears very similar to that in *F. × ananassa* June-bearing varieties, so that cool temperatures (10–16°C) or short days induce flowering, apparently by inactivating the dominant allele that represses flowering in non-inductive conditions (Brown, 1956; Brown and Wareing, 1965). We have confirmed these observations, and find that runner production shows the expected, opposite pattern to flowering (Fig. 10.15).

Conclusions

The immediate need is to understand the physiological responses of the strawberry to its environment sufficiently well to allow prediction of growth and cropping patterns. We have shown here how we are making progress towards this objective. In the longer term, there is potential for complementary methods of control, if the molecular mechanisms underlying the switches between flowering and vegetative growth can be unravelled. *F. vesca* appears to offer the best chance for progress in this area because of its genetic and developmental make up. Isolation and functional characterization of the seasonal flowering and runnering alleles will be of great physiological interest (is there an inhibitor of flowering, and if so how does it work?), and will be of practical value because of

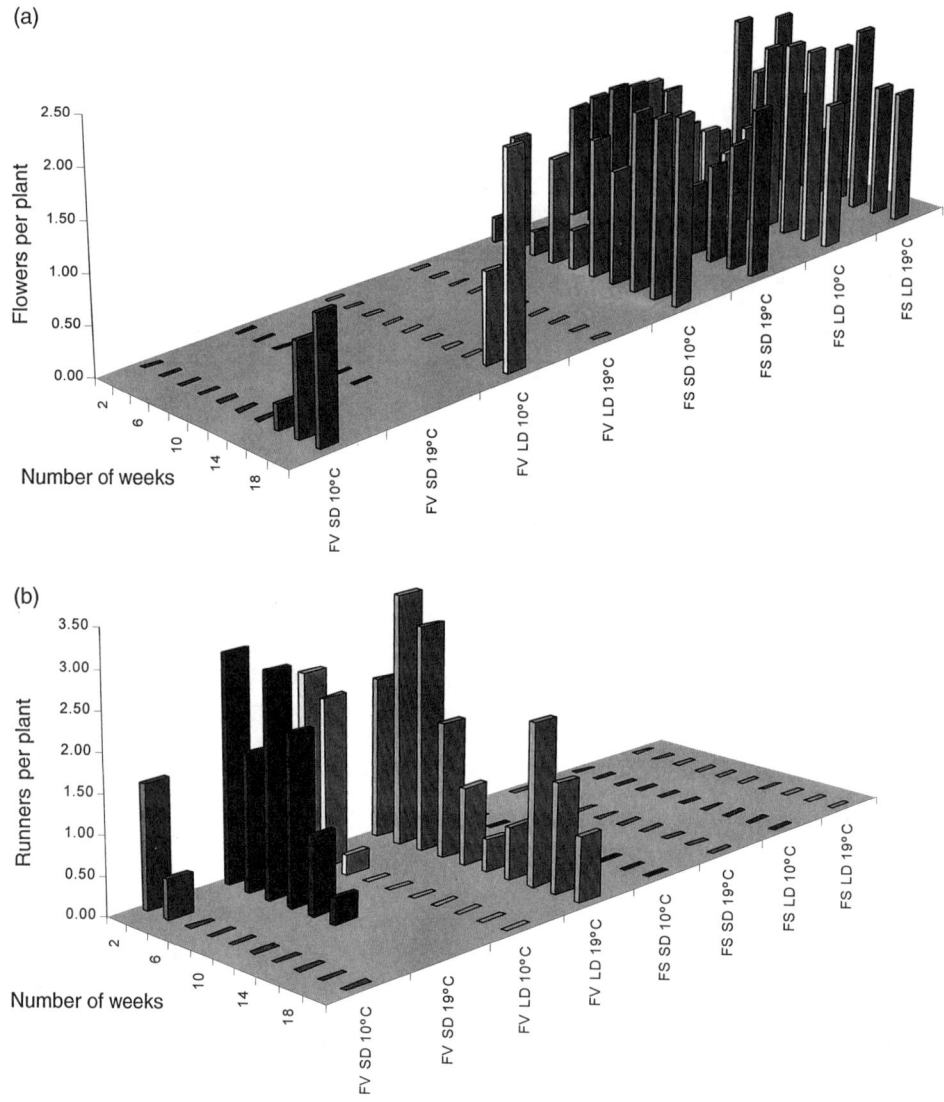

Fig. 10.15. The control of (a) flowering and (b) runnering in *F. vesca*. In *F. vesca* (FV), cool temperature (10°C) and short day (SD – 8 h) conditions induce flowering; warm temperature (20°C) and long day (LD – 17 h) conditions cause runnering. In this experiment, the *F. vesca* SD/19°C treatment was terminated after week 14 because of problems with plant health. In *F. vesca semperflorens* (FS), flowering occurs continuously and runners are not formed.

developing technologies for directed gene expression in plants. It is also likely to enhance our knowledge of perennial flowering in a wide range of species, many of which are important commercially but are only poorly understood at the molecular and physiological levels.

Acknowledgements

This paper is dedicated to Chester Guttridge. For useful discussions and advice we are grateful to Dr David Simpson, Dr David Taylor (HRI East Malling) and Dr John Jones (The University of Reading). Financial support has been provided by The University of Reading Research Endowment Trust Fund, Darby Brothers Farms Limited and The Soft Fruit Technology Group, and ZENECA Agrochemicals. A.T. is grateful to the Islamic Republic of Iran, and C.C. to Gaziosmanpasa University, Turkey for financial support. We are very grateful to Muluneh Tamiru for collecting the data shown in Fig. 10.11.

References

Ahmadi, H., Bringhurst, R.S. and Voth, V. (1990) Modes of inheritance of photoperiodism in *Fragaria*. *Journal of the American Society for Horticultural Science* 115, 146–152.

Avigdori-Avidov, H., Goldschmidt, E.E. and Kedar, N. (1977) Involvement of endogenous gibberellin in the chilling requirements of strawberry (*Fragaria* × *ananassa* Duch.). *Annals of Botany* 41, 927–936.

Bailey, J.S. and Rossi, A.W. (1965) Effects of fall chilling, forcing temperature and daylength on the growth and flowering of Catskill strawberry plants. *Proceedings of the American Society for Horticultural Science* 87, 245–252.

Bernier, G., Kinet, J.-M. and Sachs, R.M. (1981) *The Physiology of Flowering*, Vols I and II. CRC Press, Boca Raton, Florida.

Bennett, M.D. and Leitch, I.J. (1995) Nuclear DNA amounts in angiosperms. *Annals of Botany* 76, 113–176.

Brown, T. (1956). Genetical and physiological aspects of flower initiation and runner production in varieties of *Fragaria vesca* L. PhD thesis, University of Manchester, Manchester, UK.

Brown, T. and Wareing, P.F. (1965) The genetical control of the everbearing habit and three other characters in varieties of *Fragaria vesca*. *Euphytica* 14, 97–112.

Charters, Y.C., Robertson, A., Wilkinson, M.J. and Ramsay, G. (1996) PCR analysis of oilseed rape cultivars (*Brassica napus* L. ssp. *oleifera*) using 5'-anchored simple sequence repeat (SSR) primers. *Theoretical and Applied Genetics* 92, 442–447.

Craig, D.L. and Brown, G.L. (1977) Influence of digging date, chilling, cultivars and culture on glasshouse strawberry production in Nova Scotia. *Canadian Journal of Plant Science* 57, 571–576.

Darrow, G.M. (1936) Interrelation of temperature and photoperiodism in the production of fruit-buds and runners in the strawberry. *Proceedings of the American Society for Horticultural Science* 34, 360–363.

Durner, E.F. and Poling, E.B. (1988) Strawberry developmental responses to photoperiod and temperature: a review. *Advances in Strawberry Production* 7, 6–13.

El Mansouri, I., Mercado, J.A., Valpuesta, V., Lopez-Aranda, J.M., Pliego-Alfaro, F. and

Quesada, M.A. (1996) Shoot regeneration and *Agrobacterium*-mediated transformation of *Fragaria vesca* L. *Plant Cell Reports* 15, 642-646.

Erez, A. and Couvillon, G.A. (1987) Characterisation of the moderate temperature effect on peach bud rest. *Journal of the American Society for Horticultural Science* 110, 579-585.

Galletta, G.S. and Bringhurst, R.S. (1990) Strawberry management. In: Galletta, G.S. and Himelrick, D.J. (eds) *Small Fruit Crop Management*. Prentice-Hall, Englewood Cliffs, New Jersey, pp. 83-156.

Gatz, C. (1997) Chemical control of gene expression. *Annual Review of Plant Physiology and Plant Molecular Biology* 48, 89-108.

Guttridge, C.G. (1956) Photoperiodic promotion of vegetative growth in the cultivated strawberry plant. *Nature* 178, 50-51.

Guttridge, C.G. (1958) The effects of winter chilling on the subsequent growth and development of the cultivated strawberry plant. *Journal of Horticultural Science* 33, 119-127.

Guttridge, C.G. (1959a) Evidence for a flower inhibitor and vegetative growth promoter in the strawberry. *Annals of Botany* 23, 351-360.

Guttridge, C.G. (1959b) Further evidence for a growth-promoting and flower inhibiting hormone in strawberry. *Annals of Botany* 23, 612-621.

Guttridge, C.G. (1985) *Fragaria × ananassa*. In: Halevy, A. (ed.), *CRC Handbook of Flowering*, Vol III. CRC Press, Boca Raton, Florida, pp. 16-33.

Heide, O.M. (1977) Photoperiod and temperature interactions in growth and flowering of strawberry. *Physiologia Plantarum* 40, 21-26.

Ito, H. and Saito, T. (1962) Studies on the flower formation in strawberry plants. I. Effects of temperature and photoperiod on the flower formation. *Tohoku Journal of Agricultural Research* 13, 191-203.

Kronenberg, H.G. and Wassenaar, L.M. (1972) Dormancy and chilling for early forcing. *Euphytica* 21, 454-459.

Lang, G.A., Early, J.D., Martin, G.C. and Darvell, R.L. (1987) Endo-, para-, and ecodormancy: physiological terminology and classification for dormancy research. *HortScience* 23, 371-377.

Le Mière, P., Hadley, P., Darby, J. and Battey, N.H. (1996) The effect of temperature and photoperiod on the rate of flower initiation and the onset of dormancy in the strawberry (*Fragaria × ananassa* Duch.). *Journal of Horticultural Science* 71, 361-371.

Le Mière, P., Hadley, P., Darby, J. and Battey, N.H. (1998) The effect of thermal environment, planting date and crown size on growth, development and yield of *Fragaria ananassa* Duch. cv. Elsanta. *Journal of Horticultural Science and Biotechnology* (in press).

Longman, K.A., Nasr, T.A.A. and Wareing, P.F. (1965) Gravimorphism in trees. 4. The effect of gravity on flowering. *Annals of Botany* 29, 459-473.

Mika, A. (1969) Effects of shoot-bending and pruning on growth and fruit bud formation in young apple trees. *Horticultural Research* 9, 93-102.

Mika, A. (1986) Physiological responses of fruit trees to pruning. *Horticultural Reviews* 8, 337-378.

Ourecky, D.K. and Slate, G.L. (1967) Behaviour of the everbearing characteristics in strawberries. *Proceedings of the American Society for Horticultural Science* 91, 236-241.

Sadowski, J., Gaubier, P., Delseny, M. and Quiros, C.F. (1996) Genetic and physical mapping in *Brassica* diploid species of a gene cluster defined in *Arabidopsis thaliana*. *Molecular and General Genetics* 251, 298-306.

Sakin, M., Hancock, J.F. and Luby, J.J. (1997) Identifying new sources of genes that determine cyclic flowering in rocky mountain populations of *Fragaria virginiana* ssp. *glauca* Staudt. *Journal of the American Society for Horticultural Science* 122, 205-210.

Schmidt, R., West, J., Love, K., Lenehan, Z., Lister, C., Thompson, H., Bouchez, D. and Dean, C. (1995) Physical map and organization of *Arabidopsis thaliana* chromosome 4. *Science* 270, 480-484.

Sedgeley, M. (1990) Flowering of deciduous perennial fruit crops. *Horticultural Reviews* 12, 223-264.

Smeets, L. and Kronenberg, H.G. (1955) Runner formation on strawberry plants in Autumn and Winter. *Euphytica* 4, 53-57.

Tamiru, M. (1995) The effect of density and initial plant size on radiation absorption, growth and yield in strawberry (*Fragaria* × *ananassa* Duch.) cv. Elsanta. MSc thesis, University of Reading, Reading, UK.

Tehranifar, A., Le Mière, P. and Battey, N.H. (1998) The effects of lifting date, chilling duration and forcing temperature on vegetative growth and fruit production in the June-bearing strawberry cultivar 'Elsanta'. *Journal of Horticultural Science and Biotechnology* (in press).

Thomas, B. and Vince-Prue, D. (1996) *Photoperiodism in Plants*, 2nd Edn. Academic Press, London.

Uratsu, S.L., Ahmadi, H., Bringhurst, R.S. and Dandekar, A.M. (1991) Relative virulence of *Agrobacterium* strains on strawberry. *HortScience* 26, 196-199.

Vince-Prue, D. and Guttridge, C.G. (1973) Floral initiation in strawberry: spectral evidence for the regulation of flowering by long-day inhibition. *Planta* 110, 165-172.

11 Manipulating the Photoperiodic Control of Plant Reproduction

S.D. Jackson and B. Thomas

Horticulture Research International, Wellesbourne, Warwick CV35 9EF, UK

Introduction

Photoperiodism, the ability of the plant to detect, and respond to, different lengths of dark and light periods, i.e. to be able to differentiate between short days and long days, is essential for the life cycle of the plant. The invariable connection between short days and winter, and long days and summer, means that photoperiod provides a robust mechanism by which plants can chart the passing of the year. Together with the plant's response to low temperatures (vernalization), which is used to differentiate between spring and autumn when the days are of a similar length, photoperiodism is used to link the various aspects of plant development with different times of the year in such a way as to enhance the chances of survival of the plant and its progeny. For example, the dropping of leaves, onset of dormancy, development of frost hardiness and tuberization are promoted by the short days and low temperatures of autumn and thus provide the plants with a means by which they can anticipate, and adopt measures to survive, the freezing temperatures of winter. On the other hand, the breaking of bud dormancy of woody plants, and flowering in many other plants is promoted by the longer days of spring and summer. In some cases, separate stages of the same developmental process may have different photoperiodic requirements, e.g. several species of *Bryophyllum* require the longer days of summer for floral induction but then shorter days of autumn for flower development. This is a temperature-independent mechanism to ensure that flowers appear in autumn.

The most widely studied photoperiodic response is flowering because of its economic importance in agriculture and horticulture. The photoperiodic requirements for most varieties of major crops are now known and enable the best variety to be chosen for a particular latitude, thus crops are now only grown in places where the growing season incorporates days of the necessary length to induce flowering. In addition, the season of a crop can be extended by manipulating lighting conditions in the greenhouse. With some species, plant breeders can obtain up to three generations of plants per year, leading to almost year-round availability, e.g. *Chrysanthemum*, originally an autumn-flowering garden plant, has now become a major horticultural crop because horticulturalists can induce flowering in all seasons simply by manipulating the length of the light period. It is now possible to time the production of certain plants to coincide with periods of high demand, e.g. the production of poinsettias for Christmas and Easter.

Flowering is, however, a complex response which is promoted in some species by long days but in other species by short days. A simpler but closely related photoperiodic response is tuberization which is promoted by short days and inhibited by long days. Tuberization, like flowering, represents a significant developmental switch from vegetative to reproductive growth and is associated with many physiological changes in addition to the formation of tubers at the tips of stolons (reviewed in Steward *et al.*, 1981). Many factors affect the induction of tuberization, the physiological age of the plant and environmental factors, such as photoperiod, temperature and nitrogen levels, probably being the most important. In some species of potato, the photoperiodic control is strict, and these species, e.g. *Solanum demissum* and some lines of *S. tuberosum* ssp. *andigena*, will only tuberize in short days and not in long days. Such potato species are responding to photoperiod in a similar manner to qualitative short-day flowering plants such as 'Maryland Mammoth' tobacco.

The length of the dark period rather than the length of the light period is, in fact, the most important factor in photoperiodic responses, a short day representing a long night and *vice versa*. Thus, long nights induce tuberization in *S. tuberosum* ssp. *andigena* and flowering in 'Maryland Mammoth', and if the timing of a long night is interrupted by a night break then tuberization and flowering are inhibited, and the plants respond as if they are in short nights (or long days). It has also been shown that a night break of red light is the most effective at inhibiting tuberization, or flowering, and that the inhibition caused by a red light night break can be reversed if a pulse of far-red light is given immediately afterwards (King and Cumming, 1972; Deitzer *et al.*, 1979; Batutis and Ewing, 1982). Such photoreversibility is a characteristic of a phytochrome response and implicates phytochrome in the photoperiodic control of flowering and tuberization. Which of the several phytochromes that have now been identified are involved in the response to photoperiod is not yet known.

The Role of Phytochrome B

It is now known that phytochrome B plays an important role in the photoperiodic control of tuberization (Jackson *et al.*, 1996). Transformed *S. tuberosum* ssp. *andigena* plants were produced that expressed part of the potato phytochrome B gene in the antisense orientation under the control of the cauliflower mosaic virus (CaMV) 35S promoter. In two independent transformants, the high levels of expression of this antisense construct resulted in a reduction in the levels of phytochrome B mRNA and, consequently, protein. The

Fig. 11.1. Wild-type (middle) and antisense phytochrome B potato plants (left and right) grown in long days. On the antisense plants, tuber formation occurs with little or no stolon formation, whereas the wild-type plants only form stolons in long photoperiods.

level of phytochrome A protein was unaffected in the transformed potato plants, indicating that the antisense effect most probably is acting specifically on phytochrome B.

The reduced levels of phytochrome B in these transformed potato plants abolished the photoperiodic control of tuberization that is normally seen in non-transformed wild-type plants, the transformants forming tubers in long days (even in continuous light), as well as in short days. In long days, wild-type plants only form stolons, which do not develop into tubers (Fig. 11.1). Tubers were formed on the antisense plants with little or no stolon formation, which may reflect a strongly induced state of these plants to tuberize (Ewing and Wareing, 1978). The fact that plants with reduced levels of phytochrome B will tuberize in long days shows that phytochrome B plays an integral role in the photoperiodic control mechanism that regulates tuberization in potato, and that it is involved in the inhibition of tuberization in non-inductive photoperiods rather than in promoting tuberization in inducing photoperiods.

We recently have shown that phytochrome B plays the same role in the photoperiodic control of flowering in short-day plants (SDP) as it does in the control of tuberization in potato. Two independent transgenic plants of the short-day 'Maryland Mammoth' tobacco have also been produced that have reduced levels of phytochrome B RNA as a result of antisense inhibition. Like the antisense phytochrome B potato plants, these tobacco plants exhibit an altered response to photoperiod. The two antisense phytochrome B tobacco plants are

Fig. 11.2. Wild-type (left) and antisense phytochrome B 'Maryland Mammoth' tobacco plants (right) grown in long days.

able to flower in long days (16-h photoperiod) whereas non-transformed 'Mammoth' plants remain vegetative (Fig. 11.2). Furthermore, the ma_3^R mutant of the SDP *Sorghum*, which previously was shown to be almost insensitive to photoperiod in its flowering response (Pao and Morgan, 1986), is now known to be a phytochrome B mutant (Childs *et al.*, 1997).

Phytochrome B is thus involved in the inhibition of both tuberization and flowering in non-inductive long-day conditions. This being the case, it may be possible to increase the levels of inhibition by increasing the levels of phytochrome B, and thus delay flowering in inducing conditions. Short-day-requiring Hicks *MM* tobacco plants have been transformed with a construct directing the overexpression of the *Arabidopsis* phytochrome B gene from the CaMV 35S promoter (Halliday *et al.*, 1997). These phytochrome B-overexpressing plants were grown together with wild-type plants in long days and then given a short 2-week inductive treatment before being replaced into long days until they flowered. The inductive treatment varied between different groups of plants in that photoperiods of different lengths were used, i.e. some plants were given a 2-week treatment of an 8-h photoperiod, others 10, 12 or 14 h, the controls just left in long days (16 h). The length of time needed for flowering was then measured against the length of the inductive treatment (8 h being the most strongly inducing treatment). When the response curve for the phytochrome B-overexpressing tobaccos is compared with that of the wild-type plants, it is clear to see that there is a shift in the response of the transgenic plants (Fig. 11.3). The phytochrome B-overexpressing plants are not induced by photoperiods that do induce flowering in wild-type plants; the critical night length, which is the

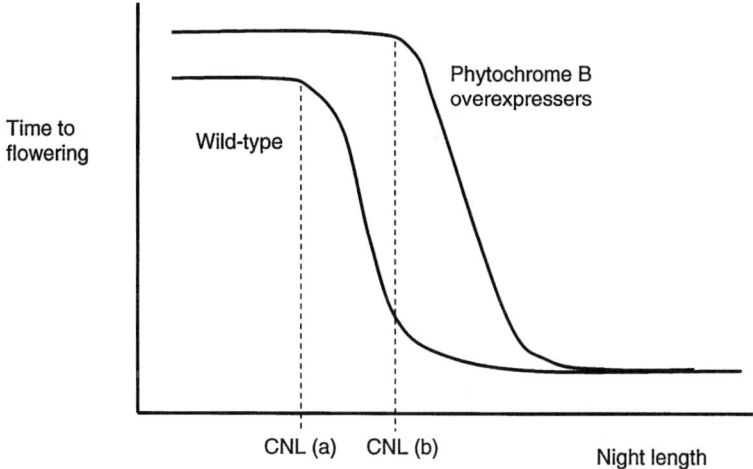

Fig. 11.3. Flowering response of wild-type and phytochrome B-overexpressing Hicks MM tobacco plants after induction by photoperiods of different lengths. The critical night length of the overexpressing tobacco plants (CNL b) is increased compared with that of wild-type plants (CNL a).

photoperiod where the response changes from non-induced to induced, of the overexpressers has been increased by around 2 h. Thus, increasing the level of phytochrome B expression alters the sensitivity of the flowering response to photoperiod, the transgenic plants requiring shorter days to induce flowering than wild-type plants.

There is also evidence that phytochrome B plays a similar role in affecting photoperiodic sensitivity in long-day plants (LDP). The *Arabidopsis phyB* mutant is early flowering compared with wild-type in both short and long days, the mutant developing fewer rosette leaves than wild-type plants before bolting (Reed *et al.*, 1993). Complementation of this mutant with a functional phytochrome B gene restores flowering time to that of the wild-type, whilst increasing the copy number of the phytochrome B gene delays flowering, i.e. they have an increased number of leaves at flowering (Wester *et al.*, 1994). So what is the mechanism by which phytochrome B modulates the sensitivity of varied responses, such as flowering and tuberization, in both SDP and LDP to photoperiod?

Inducing and Inhibitory Signals

Photoperiodic perception occurs in the leaf. Therefore, some sort of signal must be produced in response to the photoperiodic stimulus which is transmitted from leaves of the plant to the underground stolons where tuber formation occurs, or, in the case of flowering, to the apical meristem. Such a signal can be transmitted across a graft union, as was demonstrated in experiments where a leaf from a potato plant that was induced to tuberize caused a non-induced stock on to which it was grafted to tuberize, even though after grafting it was maintained in non-inducing conditions (Gregory, 1956; Chapman, 1958). Furthermore, the signal produced in leaves of tobacco plants that are induced to flower is similar to, or the same as, the signal that induces tuberization in potato. Grafting leaves from tobacco plants that were induced to flower on to potato plants that were maintained in non-inducing conditions led to tuberization of the potato plants, whereas leaves from non-induced tobacco plants did not cause tuberization (Chailakhyan et al., 1981; Martin et al., 1982). Thus, the processes leading to the production of this signal in response to an inducing photoperiod are similar in potato and tobacco, for tuberization and flowering respectively.

Such a signal may have two components, a promoting substance that increases under inductive conditions, and an inhibitory substance that decreases under inductive conditions. Grafting experiments between long-day and day-neutral tobacco species (Lang et al., 1977), and between different flowering time mutants of pea (Taylor and Murfet, 1996), have identified the existence of a floral inhibitor or 'antiflorigen'. There appears to be a direct relationship between the inducing signal and an inhibitor in tobacco. When shoots with differing numbers of leaves from the SDP 'Maryland Mammoth' and the LDP *Nicotiana sylvestris* were grafted together on to a receptor plant and kept in short days, flowering of the receptor was advanced with increasing numbers of leaves of the SDP, and delayed with increasing numbers of leaves of the LDP (Lang, 1980). The response to day length thus appears to be determined by the relative levels of inducing and inhibitory substances, the levels of at least one of these substances being affected by photoperiod.

As mentioned above, phytochrome B is involved in the inhibition of tuberization in long days rather than the induction of tuberization in short days, as removal of phytochrome B results in tuberization in both long and short days. Tuberization of the antisense phytochrome B plants in long days could be brought about by a reduction in the levels of an inhibitor, or by the production of a promoting substance, in normally non-inducing long days. That phytochrome B is involved in the production of a transmissible signal(s) has been shown by grafting experiments. A wild-type *S. tuberosum* ssp. *andigena* plant could be induced to tuberize in long days by grafting on a shoot from an antisense phytochrome B plant, but not by a graft from another wild-type plant (Jackson et al., 1998). Tuberization of such graftings does not occur, however, if some leaves are left on the wild-type stock plant. Furthermore, with the reciprocal grafting of a wild-type shoot grafted on to an antisense phytochrome B plant, tuberization of the antisense plant that would normally occur in long days is inhibited by the wild-type graft. These results indicate that an inhibitor of tuberization exists in the leaves of wild-type potato plants in long days and that the lower levels of phytochrome B in the antisense plants have led to reduced levels of this inhibitor, thus allowing tuberization to occur in long days. So what could this inhibitor be?

A Role for Gibberellins?

Certain steps in the GA (gibberellin) biosynthetic pathway are affected by photoperiod, as has been shown in spinach and pea. In spinach, bolting is prevented in short days by a lower activity of GA 20-oxidase, which results in less GA_{20} and GA_1 (Gilmour et al., 1986). In pea, senescence is prevented in short days by an increased production of GA_{53} from GA_{12} aldehyde (Davies et al., 1986). Furthermore, in many species, levels of GAs are higher in long days than in short days. It has been shown, for example, that levels of GA-like activity decrease in leaves of S. tuberosum ssp. andigena plants upon transfer from long-day to short-day conditions (Railton and Wareing, 1973).

GAs inhibit tuberization and may, therefore, play a role in the photoperiodic control of tuberization by preventing tuberization in long days. A dwarf mutant of S. tuberosum ssp. andigena that is able to tuberize in long days as well as short days has been shown to have a partial block in its GA biosynthetic pathway (van den Berg et al., 1995). Furthermore, wild-type S. tuberosum ssp. andigena plants treated with ancymidol, an inhibitor of GA biosynthesis, will tuberize in long days (Jackson and Prat, 1996). This ancymidol treatment of wild-type plants resulted in sessile tuber formation, with little or no stolon formation, in a manner very similar to the formation of tubers on the antisense phytochrome B plants. These results indicate that a decrease in GA levels may play a role in the photoperiodic induction of tuberization in potato. Also, the reduced levels of phytochrome B in the antisense plants may lead to reduced levels of, or sensitivity to GAs, therefore enabling them to tuberize in long days.

Whilst this may appear to contradict reports of increased GA levels or sensitivity in phytochrome B mutants such as the Brassica ein, Sorghum ma_3^R and cucumber lh mutants (Devlin et al., 1992; Foster et al., 1994; Lopez-Juez et al., 1995), there is a range of different biologically active GAs, with particular ones affecting different responses. It is known that the 3ß-hydroxylated gibberellin GA_1 is the most active GA with respect to stem elongation, and there is evidence from Lolium that non-3ß-hydroxylated GAs are more active in promoting flowering than 3ß-hydroxylated GAs (Evans et al., 1994). Phytochrome B may not, therefore, be causing a general reduction in the levels of all GAs, but only in specific ones, which would alter the relative levels of different GA species. By affecting the expression or activity of one or more enzymes involved in the GA biosynthetic pathway, phytochrome B could, for example, change the ratio of 3ß-hydroxylated to non-3ß-hydroxylated GAs, and thus change the development of the plant away from stem elongation and vegetative growth towards flowering and reproductive growth.

Phytochrome B does affect the level of expression of at least one GA biosynthetic enzyme, the GA 20-oxidase which is involved in the production of GA_{20} which is the immediate precursor to GA_1 before the 3ß-hydroxylation step (Fig. 11.4). The expression levels of this enzyme in the leaf are much higher throughout the plant in the phytochrome B antisense plants compared with wild-type plants. It is also interesting to note that the GA 20-oxidase catalyses the step that has been shown to be under photoperiodic control in spinach (Gilmour et al., 1986). Whether the reduced sensitivity to photoperiod in the

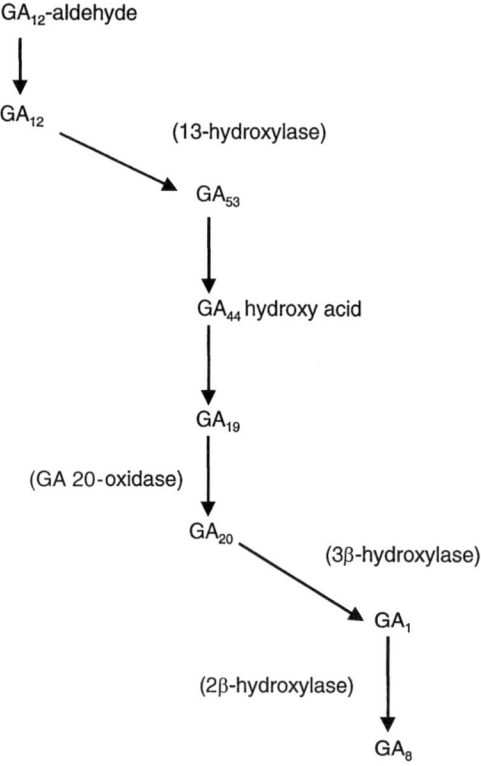

Fig. 11.4. The early 13-hydroxylation pathway which is the predominant gibberellin biosynthetic pathway in potato (van den Berg et al., 1995). The enzymes catalysing specific steps are shown in parentheses.

phytochrome B antisense potato and tobacco plants is due to the elevated GA 20-oxidase expression levels, or due to some other effect of the reduced phytochrome B levels, remains to be seen. If this proves to be the case, however, then manipulation of the GA 20-oxidase levels may be an alternative route to altering phytochrome B levels in order to modify the photoperiodic sensitivity of a plant.

Summary

The induction of tuberization and flowering in SDP are very similar photoperiodic responses that appear to act through the same mechanism. Both flowering and tuberization are initiated when a transmissible signal reaches the meristem from the leaves where it is produced in inducing photoperiods. Evidence from grafting experiments suggests that this signal has at least two components, a promoting signal and an inhibitor, and that at least one of these components is the same in both the flowering and tuberization processes. Furthermore, phytochrome B is involved in the production of one of the components of the signal controlling both flowering and tuberization. Reducing the levels of phytochrome B abolishes the photoperiodic control of tuberization in potato and of flowering in tobacco. Increasing the levels of phytochrome B has the opposite effect and increases the critical night length, meaning that shorter days are required for flowering by the phytochrome B-overexpressing plants than for wild-type plants.

Grafting experiments have shown that an inhibitor is present in the leaves of wild-type potato plants in long days that is absent in the antisense phytochrome B plants. It appears, therefore, that phytochrome B is involved in the production of this inhibitor, the levels of which determine the sensitivity of the flowering or tuberization response to photoperiod.

Although the identities of the inducing and inhibitory substances are still unknown, evidence is accumulating that certain GAs are involved as inhibitors. The fact that phytochrome B affects the levels of expression of a GA biosynthetic enzyme, the activity of which is already known to be affected by photoperiod, serves to support this theory. The manipulation of phytochrome B levels, and possibly also those of the GA 20-oxidase, is one approach that may be taken to modify the photoperiodic requirements, and therefore the harvest season, of crop plants.

References

Batutis, E.J. and Ewing, E.E. (1982) Far-red reversal of red light effect during long-night induction of potato (*Solanum tuberosum* L.) tuberisation. *Plant Physiology* 69, 672–674.

Chailakhyan, M.Kh., Yanina, L.I., Devedzhyan, A.G. and Lotova, G.N. (1981) Photoperiodism and tuber formation in grafting of tobacco on to potato. *Doklady Akademiya Nauk SSSR* 257, 1276–1280.

Chapman, H.W. (1958) Tuberisation in the potato plant. *Physiologia Plantarum* 11, 215–244.

Childs, K.L., Miller, F.R., Cordonnier-Pratt, M.-M., Pratt, L.H., Morgan, P.W. and Mullet, J.E. (1997) The *Sorghum bicolor* photoperiod sensitivity gene, Ma_3, encodes a phytochrome B. *Plant Physiology* 113, 611–619.

Davies, P.J., Birnberg, P.R., Maki, S.L. and Brenner, M.L. (1986) Photoperiod modification of (^{14}C)gibberellin A_{12} aldehyde metabolism in shoots of pea, line $G2^1$. *Plant Physiology* 81, 991–996.

Deitzer, G.F., Hayes, R. and Jabben, M. (1979) Kinetics and time dependence of the effect of far-red light on the photoperiodic induction of flowering in winter barley. *Plant Physiology* 64, 1015–1021.

Devlin, P.F., Rood, S.B., Somers, D.E., Quail, P.H. and Whitelam, G.C. (1992) Photophysiology of the elongated internode (*ein*) mutant of *Brassica rapa*. *Plant Physiology* 100, 1442–1447.

Evans, L.T., King, R.W., Mander, L.N. and Pharis, R.P. (1994) The relative significance for stem elongation and flowering in *Lolium temulentum* of 3ß-hydroxylation of gibberellins. *Planta* 192, 130–136.

Ewing, E.E. and Wareing, P.F. (1978) Shoot, stolon and tuber formation on potato

(*Solanum tuberosum* L.) cuttings in response to photoperiod. *Plant Physiology* 61, 348-353

Foster, K.R., Miller, F.R., Childs, K.L. and Morgan, P.W. (1994) Genetic regulation of development in *Sorghum bicolor*. *Plant Physiology* 105, 941-948.

Gilmour, S.J., Zeevart, J.A.D., Schwenen, L. and Graebe, J.E. (1986) Gibberellin metabolism in cell-free extracts from spinach leaves in relation to photoperiod. *Plant Physiology* 82, 190-195.

Gregory, L.E. (1956) Some factors for tuberisation in the potato. *Annual Review of Botany* 43, 281-288.

Halliday, K., Thomas, B. and Whitelam, G. (1997) Expression of heterologouos phytochromes A, B or C in transgenic tobacco plants alters vegetative development and flowering time. *The Plant Journal* 12, 1079-1090.

Jackson, S. and Prat, S. (1996). Control of tuberisation in potato by gibberellins and phytochrome B. *Physiologia Plantarum* 98, 407-412.

Jackson, S., Heyer, A., Dietze, J. and Prat, S. (1996) Phytochrome B mediates the photoperiodic control of tuber formation in potato. *The Plant Journal* 9,159-166.

Jackson, S., James, P., Prat, S. and Thomas B. (1998) Phytochrome B affects the levels of a graft-transmissible signal involved in tuberisation. *Plant Physiology* 117, 29-32.

King, R.W. and Cumming, B.G. (1972) The role of phytochrome in photoperiodic time measurement and its relation to rhythmic timekeeping in the control of flowering in *Chenopodium rubrum*. *Planta* 108, 39-57.

Lang, A. (1980) Inhibition of flowering in long-day plants. In: Skoog, F. (ed.) *Plant Growth Substances 1979*. Springer-Verlag, Berlin, pp. 310-322.

Lang, A., Chailakhyan, Kh. and Frolova, I.A. (1977) Promotion and inhibition of flower formation in a day-neutral plant in grafts with a short-day plant and a long-day plant. *Proceedings of the National Academy of Sciences USA* 74, 2412-2416

Lopez-Juez, E., Kobayashi, M., Sakurai, A., Kamiya, Y. and Kendrick, R.E. (1995) Phytochrome, gibberellins and hypocotyl growth. *Plant Physiology* 107, 131-140.

Martin, C., Vernay, R. and Paynot, N. (1982) Physiologie vegetale. Photoperiodisme, tuberisation, floraison et phenolamides. *Comptes Rendus Hebdomadaires des Seances de l'Academie des Sciences* 295, 565-568.

Pao, C-I. and Morgan, P.W. (1986) Genetic regulation of development in *Sorghum bicolor*. *Plant Physiology* 82, 581-584.

Railton, I.D. and Wareing, P.F. (1973) Effects of daylength on endogenous gibberellin levels in leaves of *Solanum andigena*. *Physiololgia Plantarum* 28, 88-94.

Reed, J.W., Nagpal, P., Poole, D.S., Furuya, M. and Chory, J. (1993) Mutations in the gene for the red/far-red light receptor phytochrome B alter cell elongation and physiological responses throughout *Arabidopsis* development. *The Plant Cell* 5, 147-157.

Steward, F.C., Moreno, V. and Roca, W.M. (1981) Growth form and composition of potato plants as affected by environment. *Annals of Botany* 48 (Suppl. 2), 45.

Taylor, S.A. and Murfet, I.C. (1996) Flowering in *Pisum*: identification of a new *ppd* allele and its physiological action as revealed by grafting. *Physiologia Plantarum* 97, 719-723.

van den Berg, J.H., Simko, I., Davies, P.J., Ewing, E.E. and Halinska, A. (1995) Morphology and (^{14}C)gibberellin A$_{12}$ aldehyde metabolism in wild-type and dwarf *Solanum tuberosum* ssp. *andigena* grown under long and short photoperiods. *Journal of Plant Physiology* 146, 467-473.

Wester, L., Somers, D.E., Clack, T. and Sharrock, R.A. (1994) Transgenic complementation of the *hy3* phytochrome B mutation and response to *PHYB* gene copy number in *Arabidopsis*. *The Plant Journal* 5, 261-272.

12 Regulation of Abscisic Acid and Water Stress Response Genes

P.K. Busk, M. Figueras, A.C. Jessop, A. Goday and M. Pagès

Departament de Genética Molecular, C.I.D. (C.S.I.C.), Jordi Girona 18–26, 08034 Barcelona, Spain

Metabolism of ABA

ABA biosynthesis

The endogenous concentration of abscisic acid (ABA) changes during development and in response to stress (reviewed by Skriver and Mundy, 1990), suggesting a tight control of ABA biosynthesis. Fungi synthesize ABA by a direct pathway involving only C15 compounds (reviewed by Zeevart and Creelman, 1988). There is no evidence for the existence of this pathway in higher plants. Instead, ABA is synthesized from mevalonic acid through an indirect pathway from the C40 carotenoid precursors zeaxanthin, violaxanthin and neoxanthin (Fig. 12.1). The reduced capability to produce ABA of the carotenoid biosynthesis mutants *vp2*, *vp5*, *vp7*, *vp9*, *w3*, *y3* and *y9* from maize shows the connection between carotenoids and ABA. Genetics has also helped to elucidate the later steps of the pathway. The *aba* mutants from *Arabidopsis* and *Nicotiana plumbaginifolia* are impaired in the conversion of zeaxanthin into antheraxanthin and fail to produce violaxanthin and neoxanthin. The responsible enzyme has been cloned and is an epoxidase capable of converting zeaxanthin into all-*trans*-violaxanthin (Marin *et al.*, 1996). The changes in the concentrations of xanthoxins and ABA during water stress suggested that all-*trans*-violaxanthin is converted into 9'-*cis*-neoxanthin through a branched pathway (Fig. 12.1). Neoxanthin is cleaved in an oxidation step to produce xanthoxin that is converted into ABA aldehyde. In water-stressed leaves, the cleavage is increased, suggesting that water stress regulates ABA synthesis by stimulating the activity of a dioxygenase necessary for cleavage of neoxanthin. Induction requires synthesis of either the dioxygenase or another protein, because cycloheximide can inhibit the increase in ABA concentration. Regulation is probably at the transcriptional level as actinomycin D is also an inhibitor (Guerrero and Mullet, 1986). Characterization of the dioxygenase is likely to

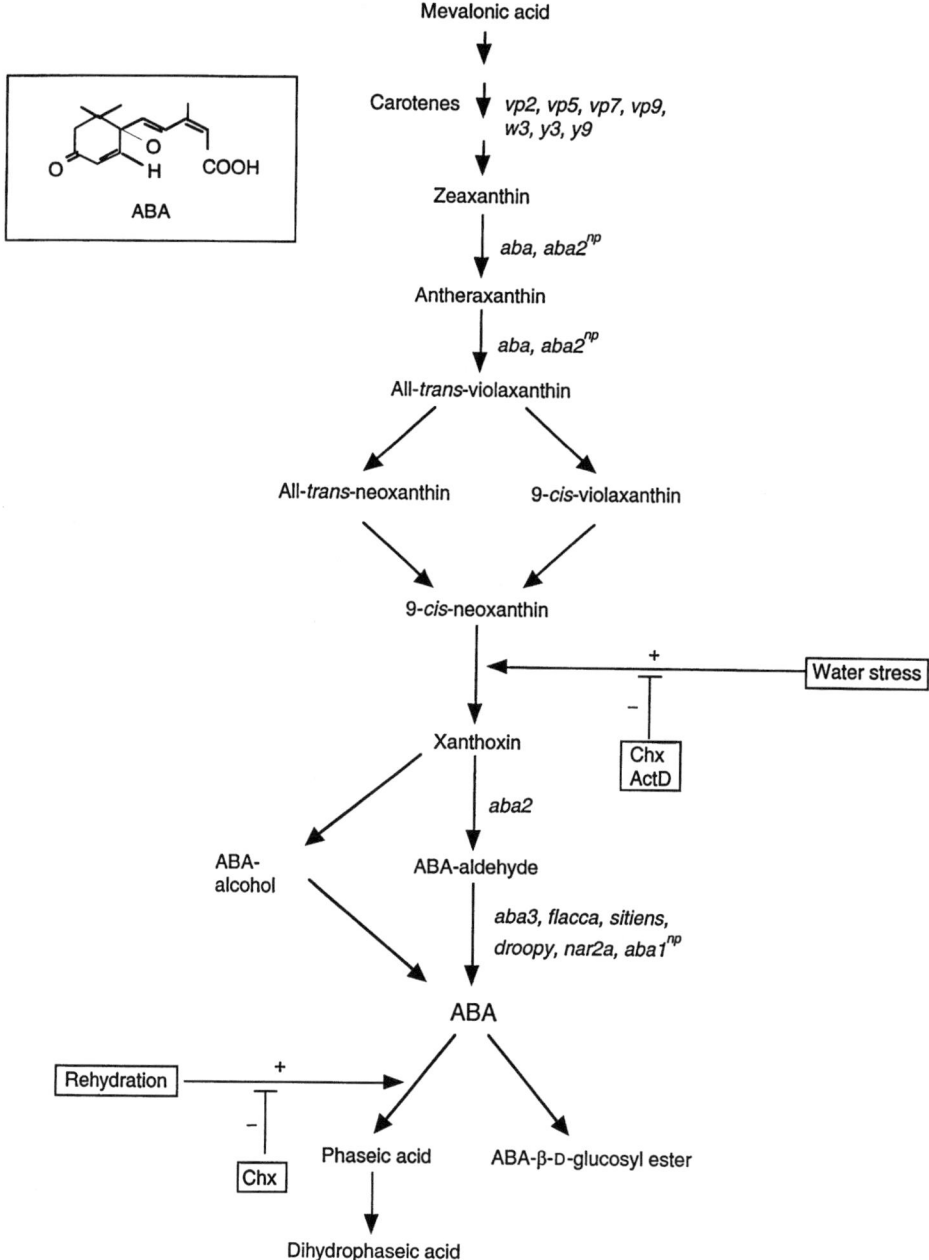

Fig. 12.1. ABA metabolism in plants. Chx, cycloheximide; ActD, actinomycin D. The metabolic mutants of each step are indicated in italics.

produce important information about the response of ABA synthesis to abiotic stress.

A recent report showed that the *Arabidopsis* mutant *aba2* is deficient in the conversion of xanthoxin into ABA aldehyde. Mutants deficient in the conversion of ABA aldehyde into ABA have been isolated from several plant species (Fig. 12.1). However, these mutants produce low but physiologically significant levels of ABA through a shunt pathway including ABA alcohol (Rock *et al.*, 1991). This pathway can be inhibited by carbon monoxide, suggesting the involvement of a P_{450} monooxygenase.

ABA catabolism

The major ABA catabolites are phaseic acid, 4'-dihydrophaseic acid and abscisic acid-ß-D-glycosyl ester (Fig. 12.1) (Zeevart and Creelman, 1988). However, there are differences between species in the catabolism of ABA. The rate of conversion of ABA into phaseic acid increases during water stress until it equals the rate of synthesis of ABA after about 7 h. Thus, a balance between synthesis and catabolism maintains the high ABA level in stressed tissues. From a regulatory point of view, it is interesting that the rate of conversion of ABA into phaseic acid increases during rehydration (reviewed by Zeevart and Creelman, 1988). Cycloheximide can inhibit the induced degradation of ABA, demonstrating that increased ABA catabolism during rehydration requires protein synthesis.

The Biological Role of ABA

Embryo dormancy, germination and desiccation tolerance

The concentration of ABA increases in late embryo development shortly before the onset of desiccation and seed dormancy (King, 1976). Genetic studies have shown that ABA is involved in dormancy, and ABA-deficient mutants have been isolated by screening for precocious germination or reversion of germination-deficient mutations (reviewed by Koornneef, 1986). Reciprocal crosses of wild-type and *aba* mutants of *Arabidopsis* showed that in this species there is a peak of ABA of maternal origin halfway through seed development and a smaller peak of embryonic ABA in later stages. Interestingly, the embryonic ABA is necessary for seed dormancy whereas maternal or applied ABA are incapable of inducing this trait. This shows that there are two different pools of ABA in *Arabidopsis* seeds. The maternal ABA has a significant role in regulating seed maturation and inhibiting precocious germination, as demonstrated by crossing *aba* mutants with the ABA-insensitive mutant *abi3* (Koornneef *et al.*, 1989). The *aba*, *abi3* mutant seeds were viviparous and desiccation intolerant when developing on a mother plant with low ABA levels (*aba* mutant), but these phenotypes were suppressed on mother plants with normal ABA levels (heterozygous for the *aba* mutation). Maternal ABA is not sufficient to induce seed maturation and repress precocious germination in maize because the ABA-deficient *viviparous* mutants germinate prematurely on the cob independently of the maternal phenotype (Robertson, 1955).

An important point of the study of Koornneef *et al.* (1989) is that seed development is much more sensitive to ABA than is induction of dormancy. Germination of wild-type, ABA-deficient and ABA-insensitive mutants was tested at different days after pollination (d.a.p.) and showed that only the wild-type undergoes dormancy. However, all the mutants except the *aba, abi3* double mutant were desiccation tolerant. This is explained by the leaky nature of the mutants. The ABA concentration in the *aba* mutant is too low to induce dormancy but sufficient to confer desiccation tolerance. Only when this mutant is combined with reduced sensitivity to ABA (*abi3*) is desiccation tolerance affected. Thus, the wild-type levels of ABA appear to control dormancy but are far too high to be the primary regulator of maturation. However, that ABA is involved in maturation is demonstrated by the phenotype of the *aba, abi3* double mutant which showed reduced water loss and desiccation tolerance, blockage of testa development and lack of storage proteins.

Galau and co-workers identified classes of coordinately expressed mRNAs in cotton embryos *in vivo* and under various culture conditions (Hughes and Galau, 1991). A few temporal programmes that are unrelated to the variations in ABA concentrations can explain the expression patterns. These programmes appear to be controlled by a maturation factor and a post-abscission factor. The model seems to be valid also in *Arabidopsis* (Giraudat *et al.*, 1994; Parcy *et al.*, 1994). The maturation and post-abscission factors have not been identified, but several mutants affecting embryo maturation are known (reviewed by McCarty, 1995). The post-abscission factor is thought to act at the same developmental stage as ABA and is, therefore, most relevant for understanding the effects of this hormone.

Growth and desiccation tolerance of vegetative tissues

ABA mediates the response of plants to drought. As described above, desiccation stimulates ABA synthesis and there are high concentrations of ABA in water-stressed leaves. Moreover, most ABA-deficient mutants display an excessive water loss due to defects in stomatal regulation which leads to an increased tendency to wilt (reviewed by Zeevart and Creelman, 1988; Giraudat *et al.*, 1994). Characterization of ABA-deficient tomato plants showed a direct relationship between stomatal closure and ABA. The stomata of these plants were unresponsive to water stress but could be closed by applying ABA. ABA synthesized in water-stressed roots may act as a chemical signal to induce stomatal closure before any desiccation of the leaves (Tardieu, 1996). When the roots are submitted to stress, they will produce ABA that is transported to the aerial parts of the plant and induces stomatal closure to prevent evaporation. This could be a mechanism to optimize the plant's water use under restricted availability.

ABA-deficient mutants are wilty, which shows a role for ABA in reducing water loss during normal growth, but ABA does not seem to be involved in slow adaptation to low water potential (Leone *et al.*, 1994). This result is based on the observation that ABA levels do not rise in potato cell cultures adapted to high polyethylene glycol (PEG) concentrations. Characterization of the response of

ABA-deficient mutants to this treatment will show if adaptation involves the basal level of ABA.

An interesting model system for studying the biological role of ABA in water stress is the resurrection plant *Craterostigma plantagineum* which can be dried to 1% relative water content and is still viable after rehydration. Callus cultures of *C. plantagineum* are only able to survive desiccation after 4 days of treatment with ABA. Thus, in this case, ABA is able to induce a slow adaptation to low water potential. Recently, the CDT-1 gene in which overexpression leads to desiccation tolerance in the absence of ABA has been isolated (Furini *et al.*, 1997). Characterization of the mechanism of action of CDT-1 will clarify the role of ABA.

Many of the maize viviparous mutants are carotenoid deficient and have phenotypes that are not related to ABA levels. However, the *vp8* mutant, which is probably blocked in a late stage of ABA biosynthesis that does not affect carotenoid levels, produces dwarfed plants (Robertson, 1955). Reduced growth of ABA-deficient mutants is not a direct effect of ABA on cell division or elongation (Koornneef, 1986). The three ABA-deficient tomato mutants *flacca*, *sitiens* and *notabilis* are dwarfish under normal conditions but growth in high humidity reverses the phenotype, showing that dwarfism is an effect of increased water loss rather than of the ABA content. Nevertheless, exogenous ABA reduces growth of aerial plant parts by inhibiting cell wall loosening. Reports on the effects of applied ABA on root growth are contradictory since ABA can both inhibit and promote growth of this organ (reviewed by Zeevart and Creelman, 1988). Furthermore, the endogenous ABA levels in roots show no correlation with growth rate. The isolation of mutants that are insensitive to ABA-mediated growth inhibition will be an important tool to clarify the role of ABA in cell growth.

ABA-induced Gene Expression

Many genes which are expressed in late embryo development can be induced by exogenous ABA treatment in embryo and vegetative tissues (Gomez *et al.*, 1988; Mundy and Chua, 1988). However, expression is not correlated with the endogenous ABA concentration, suggesting that other factors regulate these genes in the embryo (Hughes and Galau, 1991; Pla *et al.*, 1991). Gene expression in late embryo development is tightly associated with desiccation tolerance. ABA seems to be a prerequisite for expression, but the wild-type level is far in excess of the necessary concentration. This is demonstrated by the levels of gene expression in mutants with low ABA levels. It is known that in the *viviparous* mutants of maize, accumulation of *rab17* (Pla *et al.*, 1989), *rab28* (Pla *et al.*, 1991), *Em* and *Cat1* (Williamson and Scandalios, 1992) is reduced but not absent. The relatively high expression in the mutants does not reflect the low level of endogenous ABA. Similar results have been found in the *Arabidopsis aba* mutant (Finkelstein, 1994). The *Em* gene is strictly seed specific whereas *rab28* is inducible by ABA and stress in vegetative tissues. The transcriptional activator VP1 is necessary for the embryo-specific expression of *Em* and *rab28*. VP1 is a seed-specific protein that controls a number of ABA-inducible genes (reviewed by McCarty, 1995). The *vp1* mutants have normal ABA levels but exhibit reduced sensitivity to exogenous ABA. The lack of expression of *Em* and *rab28* in the *vp1* mutant seems to be caused by a reduced sensitivity of the genes to ABA. In favour of this, applied ABA induces both genes in excised *vp1* embryos (Pla et al., 1991; Williamson and Scandalios, 1992). The developmental expression of VP1 does not correlate with the expression of *Em* and *rab28* (McCarty, 1995; Niogret et al., 1996). VP1 is present before the induction of *Em* and *rab28* but seems to be a prerequisite for induction of these genes. The lack of correlation in the expression pattern indicates that VP1 is not the only developmental factor regulating expression in response to ABA. In agreement with this, VP1 is not required for expression of the ABA-responsive *Cat1* gene which is active during late embryo development like *Em* and *rab28* (Williamson and Scandalios, 1992). Expression of the *rab17* gene also seems to be VP1 independent (M.F. Niogret and M. Pagès, unpublished).

A VP1 homologue, ABI3, from *Arabidopsis* has been cloned, showing the conservation of this gene between monocot and dicot species (Giraudat et al., 1994). ABI3 has many of the same functions as VP1 and regulates expression of ABA-responsive genes during late embryo development (Parcy et al., 1994), e.g. mRNAs of the *Em* homologues *AtEM1* and *AtEM6* are absent in an *abi3* mutant. Transgenic *Arabidopsis* expressing ABI3 in vegetative tissues results in the induction of otherwise seed-specific genes in response to ABA. Interestingly, *AtEM1* but not *AtEM6* is ABA responsive in the transgenic plants, suggesting that seed-specific factors other than ABI3 are necessary for expression of *AtEm6*. ABI3 and VP1 have many ABA-independent functions and are therefore not an integrated part of ABA signalling (McCarty, 1995). However, the severe effects on gene expression of double mutants of *abi3* with other ABA-insensitive mutants (*abi1* and *abi2*) show that ABI3 interacts with the ABA signal pathways to regulate ABA-inducible genes. Therefore, ABI3/VP1 and ABA are parts of different pathways that stimulate each other by inducing common target genes.

ABA Signal Pathways

All well-characterized ABA-deficient mutants encode ABA biosynthetic enzymes. No regulatory mutant affecting the timing or organ specificity of ABA synthesis has been found. The problem is complicated, at least in dicotyledons, by the significant role of maternal ABA in embryo development. Dormancy is induced by embryonic ABA, but screening for altered timing of dormancy has only yielded ABA-insensitive mutants or mutants of ABA-independent pathways (Léon-Kloosterziel et al., 1996) apart from ABA biosynthetic mutants. Characterization of the regulation of the ABA biosynthetic enzymes will probably yield information about regulation of hormone synthesis. However, no seed-specific ABA-deficient mutants are known, although the maize *vp8* and *vp10* are possible candidates (McCarty, 1995). Surprisingly, the mRNA of the ABA biosynthetic enzyme, zeaxanthin epoxidase, is not detectable in embryos although the corresponding mutants (*Arabidopsis aba* and *N. plumbaginifolia aba2*) have an ABA-deficient phenotype in seed (Koornneef *et al.*, 1986; Marin *et al.*, 1996).

It is thought that desiccation and other stresses leading to low water potential result in cell wall loosening (reviewed by Zeevart and Creelman, 1988; Bray, 1993). This in turn induces a signalling cascade that includes synthesis of ABA, through a pathway that requires protein synthesis (Fig. 12.1). Despite the evidence for several ABA receptors, none has been cloned.

Intracellular signalling proteins

Several recent studies have shown that kinase and phosphatase inhibitors affect gene expression in response to ABA, and the finding that ABI1 and ABI2 encode protein phosphatases provides a genetic demonstration of the involvement of phosphorylation events in ABA signalling. Urao *et al.* (1994) cloned two closely related Ca^{2+}-dependent protein kinases from *Arabidopsis* and showed that they are ABA inducible. Overexpression of these kinases in protoplasts isolated from maize leaves induces an ABA-responsive promoter. The ABA signal pathway that includes Ca^{2+}-dependent protein kinases is probably activated by a rise in intracellular Ca^{2+} levels (Sheen, 1996). In favour of this, kinase antagonists inhibit Ca^{2+}-dependent gene expression in response to ABA. The inhibition is on a kinase downstream of Ca^{2+} in the signal pathway. Ca^{2+}-binding calmodulin proteins have been implicated in Ca^{2+} signalling in plants, but the Ca^{2+}-dependent protein kinases activated by ABA possess a calmodulin-like domain (Urao *et al.*, 1994). These kinases can therefore respond directly to Ca^{2+} without the need for calmodulin.

Also, mitogen-activated protein kinases (MAPKs) are involved in ABA signal transduction. MAPKs are part of signal cascades involving several phosphorylation events and are themselves activated by phosphorylation (reviewed by Jonak *et al.*, 1994). The MAPK family is highly conserved between species as different as mammals, yeast and plants. Application of ABA to barley aleurone protoplasts activates a MAPK by phosphorylation. Induction is rapid and MAPK activity peaks 3 min after addition of ABA. The use of a phosphatase inhibitor showed that dephosphorylation by a tyrosine phosphatase (not ABI1) is necessary for activation of MAPK. The same phosphatase inhibitor blocks induction of the *rab16* gene by ABA in aleurone (Heimovaara-Dijkstra *et al.*, 1995). Also, ABA-induced expression of *HVA1* in aleurone requires dephosphorylation. A likely pathway for induction of gene expression by ABA includes dephosphorylation of a MAPK kinase which, in turn, activates MAPK by phosphorylation. As discussed, ABA-induced gene expression does not involve Ca^{2+} in the aleurone and is probably mediated by changes in pH. Therefore, the MAPK pathway leading to gene expression in response to ABA is likely to be activated by a change in intracellular pH in these cells.

Promoter analysis

An important step is the characterization of the *cis*-acting elements mediating ABA and stress response. Functional dissection of promoters of ABA-responsive genes has revealed the existence of a *cis*-acting ABA-responsive element (ABRE) containing the core sequence ACGT which confers ABA responsiveness upon a reporter gene (reviewed by Skriver and Mundy, 1990). Several proteins that bind to the ABRE *in vitro* have been cloned, but none has been shown to regulate ABA-inducible transcription *in vivo*. In addition, elements with the same sequence as the ABRE are involved in regulation of ABA-independent genes. It has been shown that the sequence flanking the ACGT core is important for the function of the ABRE *in vivo* (Salinas *et al.*, 1992) and *in vitro* (Izawa *et al.*, 1993). A 35S minimal promoter with six copies of the sequence GTACGTGGCGC is induced by ABA, showing that, in this case, the ABRE is sufficient for ABA response. However, the two ABREs in the barley *HVA22* gene can only confer induction by ABA in the presence of additional sequences from the promoter. It was proposed that, in general, the specificity for ABA response is determined by the ABRE and a coupling element. One coupling element was isolated and shown to have the sequence TGCCACCGG; furthermore, there is a synergistic effect between ABRE and other elements in ABA-induced transcription (Shen *et al.*, 1996). Also, in the *CDeT27-45* gene from *C. plantagineum*, the activity of the ABREs is dependent on other elements in the promoter (Nelson *et al.*, 1994).

An interesting element is the DRE (drought-responsive element) which mediates drought-inducible transcription but not ABA-inducible transcription in *Arabidopsis* (Yamaguchi-Shinozaki and Shinozaki, 1994). These data suggest that several elements participate in the regulation of drought- and ABA-inducible genes.

Regulation of *rab17* and *rab28* genes from maize

In maize, *rab17* and *rab28* are typical ABA-inducible genes and are expressed both in the embryo and during drought, and in response to applied ABA. Characterization of the *rab* genes in ABA-deficient maize and in transgenic *Arabidopsis* showed differential regulation in embryos and in vegetative tissues (Pla *et al.*, 1991; Vilardell *et al.*, 1994). In addition, different proteins bind to an ABRE from *rab28 in vitro*, suggesting that tissue-specific factors regulate *rab28* even though the gene is ABA inducible in all cell types investigated (Pla *et al.*, 1993; Niogret *et al.*, 1996). VP1 is a seed-specific transcription factor that is necessary for expression of *rab28* in the embryo. Results from transient transformation have shown that VP1 and ABA have a synergistic effect on induction of transcription and that the ABRE is necessary for this effect (Hattori *et al.*, 1995). VP1 stimulates binding of the transcription factor EMBP-1 to the ABRE *in vitro* (Hill *et al.*, 1996). However, it is not known how VP1 functions *in vivo*.

Different elements regulate *rab17* in embryos and in leaves

By *in vivo* footprinting, nine elements that bind proteins have been identified in the *rab17* promoter (Busk *et al.*, 1997). The footprints were different in embryos and leaves, suggesting that different proteins bind to the elements in the two tissues. Six of the elements were important for transcription in transient transformation of embryos, whereas only three elements were important in leaves of 3-day-old seedlings. The elements can be divided into embryo-specific, ABA-specific and leaf-specific sequences depending on the footprint *in vivo* and the role in induction of *rab17*. These results show that an embryo-specific pathway regulates *rab17* in embryos while another pathway is important for induction by drought and ABA in vegetative tissues. This provides a molecular basis for the differential regulation of the *rab* genes in different tissues (Pla *et al.*, 1991; Vilardell *et al.*, 1994).

Five of the characterized elements are putative ABREs, but not all of these were important for regulation of *rab17*. Probably, different proteins bind to the ABREs in embryos and in leaves (Pla *et al.*, 1993). Furthermore, there was a high degree of cooperation between the elements: if one element was mutated, the activity of the promoter decreased drastically, especially in embryos. These results suggest that the function of the ABREs is determined by the proteins that bind to the elements and by influences from other elements. It previously has been shown that two elements can cooperate in ABA-induced transcription (Shen *et al.*, 1996). In the *rab17* promoter, several elements cooperate.

The four non-ABRE elements in the *rab17* promoter are two DRE-like elements, an Sph element and a new element called GRA for GC-rich Rab activator. GRA is an element that has not been described previously. The element has different footprints and activities in embryos and leaves and is therefore interesting for studying developmental regulation. In contrast to the DRE from *Arabidopsis* (Yamaguchi-Shinozaki and Shinozaki, 1994), the DRE2 from *rab17* is regulated by ABA. This could be an important difference between monocotyledons and dicotyledons and should be investigated by characterization of additional DRE elements from maize or other monocotyledons. A protein that binds to the DRE has been cloned from *Arabidopsis* (Stockinger *et al.*, 1997). It would be interesting to know if a homologous protein exists in maize.

Biotechnological Applications

The implication of ABA in many biological processes makes ABA-related genes interesting for biotechnology. The resistance of seeds to storage (dormancy) and desiccation, and stress tolerance of vegetative tissues are obvious targets for modifications. The discussion below focuses on desiccation tolerance but is also valid for the modification of other ABA-related biological processes.

Many ABA-inducible genes encode proteins that are believed to protect the plant against the effects of dehydration, cold and high salt (reviewed by Bohnert *et al.*, 1995; Close, 1996). These genes are interesting for attempts to increase the plant's resistance to stress conditions. Indeed, in some cases, transgenic plants that overexpress ABA-inducible genes are more resistant to stress than wild-type plants (M. Figueras, A. Jessop and M. Pagès, unpublished). However, adaptation to stress involves changes in many physiological processes, and it is unlikely that one gene product will be sufficient to produce more than a marginal increase in stress tolerance. Therefore, several genes will have to be expressed in a transgenic plant to achieve biotechnologically useful effects.

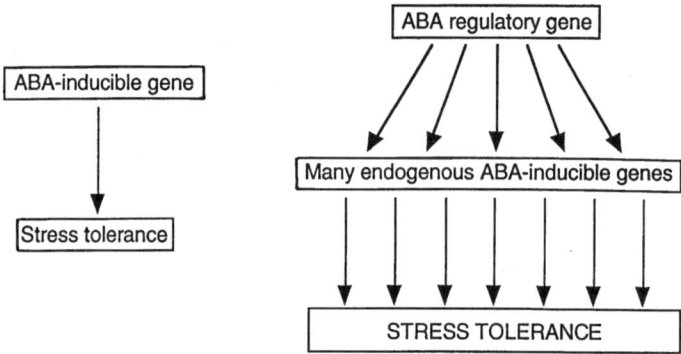

Fig. 12.2. Two strategies for increasing stress tolerance in plants.

Knowledge of the regulation of ABA-inducible gene expression can help to increase the stress resistance of plants. High expression of several genes involved in stress tolerance can be achieved by making transgenic plants with a single regulatory gene such as a protein kinase, phosphatase or a transcription factor of the ABA signal pathways (Fig. 12.2). In principle, any regulatory protein from the ABA receptors to the ABA-induced transcription factors are interesting in this context. Embryo-specific genes are inducible by stress in leaves of transgenic *Arabidopsis* with ectopic expression of the *ABI3* gene (Parcy *et al.*, 1994). These embryo-specific genes are believed to confer desiccation tolerance to the embryo and could possibly improve the stress tolerance of vegetative tissues. Also the genes that control ABA metabolism are interesting for improving stress tolerance. For example, overexpression of enzymes involved in ABA synthesis could lead to higher induction of genes that confer tolerance.

However, problems can also arise if the genetically modified plant or products are used for human consumption. Transgenic tobacco with the mannitol 1-phosphate dehydrogenase gene accumulates mannitol and has improved drought tolerance. However, alteration of sugar metabolism may change the taste of the edible plant parts. Tissue-specific promoters can be used to direct expression to the relevant plant part. For example, transgenic maize with the mannitol 1-phosphate dehydrogenase gene under control of a promoter that is only active in vegetative tissues could produce plants with increased desiccation tolerance. As the gene would not be expressed in seeds, the taste of the seeds, which are for human consumption, would not be altered.

Understanding of ABA-related processes can also help in traditional breeding. One possibility is to screen plants for overexpression of ABA regulatory genes. Screening for gene expression is a much faster selection procedure than phenotypic selection for increased desiccation resistance. The recent characterization of ABA-hypersensitive mutants is interesting in this context. ABA-inducible promoters could be used for expressing genes that are not related to stress resistance. The construction of artificial promoters provides a means to direct expression to the tissue and conditions of interest (Shen *et al.*, 1996). One use of this approach could be to express lysine-rich proteins under the control of ABA-inducible, seed-specific promoters in maize. The maize seed has low levels of lysine. Therefore, enrichment for this amino acid would improve the nutritional value of the seed.

Summary

The phytohormone ABA plays an important role in many physiological processes in plants. This hormone is necessary for regulation of several events during late seed development (reviewed by McCarty, 1995). Furthermore, ABA is crucial for the response to environmental stresses such as desiccation, salt and cold (reviewed by Bray, 1993; Jensen et al., 1996).

Recently, important advances have been made in understanding the pathways that induce ABA and how the ABA signal is transduced into physiological responses. In this context, the regulation of the biosynthesis of ABA is central. The recent cloning of a gene encoding an enzyme involved in the synthesis of ABA is a valuable tool to improve our understanding of the regulation of ABA biosynthesis (Marin et al., 1996).

The study of gene expression in response to ABA and stress has led to the detection of several intermediates of the ABA signal cascade. The results have been achieved mainly by the use of possible intermediates and antagonists of the pathway and by examining the defects in ABA-deficient or insensitive mutants (reviewed by Giraudat et al., 1994).

ABA induces the transcription of many genes, and dissection of ABA-responsive promoters has given new insight into the integration of ABA into stress response and seed development. Several ABA- and stress-responsive *cis*-elements have been identified, and the interactions between these elements are being investigated (reviewed by Skriver and Mundy, 1990; Busk and Pagès, 1998).

Efforts have also turned toward understanding functions of proteins that are induced during desiccation. Proteins accumulated during stress are anticipated to promote cellular tolerance of dehydration through protective functions in the nucleus and cytoplasm, alteration of cellular water potential to promote water uptake, control of ion accumulation and further regulation of gene expression. Among them the Rabs from maize (Vilardell et al., 1990; Goday et al., 1994; Niogret et al., 1996) have been largely characterized.

Acknowledgements

This work was supported in part by the European Communities Biotechnology programme BIO4-CT96-0062 and in part by grant BIO97-1384 E from Plan Nacional de Investigación Científica y Desarrollo Tecnológico.

References

Bohnert, H.J., Nelson, D.E. and Jensen, R.G. (1995) Adaptations to environmental stresses. *The Plant Cell* 7, 1099-1111.

Bray, E.A. (1993) Molecular responses to water deficit. *Plant Physiology* 103, 1035-1040.

Busk, P.K. and Pagès, M. (1998) Regulation of abscisic acid-induced transcription. *Plant Molecular Biology* 37, 425-435.

Busk, P.K., Jensen, A.B. and Pagès, M. (1997) Regulatory elements *in vivo* in the promoter of the abscisic acid responsive gene *rab17* from maize. *The Plant Journal* 11, 1285-1295.

Close, T.J. (1996) Dehydrins: emergence of a biochemical role of a family of plant dehydration proteins. *Physiologia Plantarum* 97, 795-803.

Finkelstein, R.R. (1994) Mutations at two new *Arabidopsis* ABA response loci are similar to the *abi3* mutations. *The Plant Journal* 5, 765-771.

Furini, A., Koncz, C., Salamini, F. and Bartels, D. (1997) High level transcription of a member of a repeated gene family confers dehydration tolerance to callus tissue of *Craterostigma plantagineum*. *EMBO Journal* 16, 3599-3608.

Giraudat, J., Parcy, F., Bertauche, N., Gosti, F., Leung, J., Morris, P.C., Bouvier-Durand, M. and Vartanian, N. (1994) Current advances in abscisic acid action and signalling. *Plant Molecular Biology* 26, 1557-1577.

Goday, A., Jensen, A., Culiañez-Macià, F.A., Albà, M.M., Figueras, M., Serratosa, J., Torrent, M. and Pagès, M. (1994) The maize abscisic acid responsive protein rab17 is located in the nucleus and cytoplasm and interacts with nuclear localization signals. *The Plant Cell* 6, 351-360.

Gomez, J., Sanchez-Martinez, D., Stiefel, V., Rigau, J., Puigdonmenech, P. and Pagès, M. (1988) A gene induced by the plant hormone abscisic acid in response to water stress encodes a glycine-rich protein. *Nature* 334, 262-264.

Guerrero, F. and Mullet, J.E. (1986) Increased abscisic acid biosynthesis during plant dehydration requires transcription. *Plant Physiology* 80, 588-591.

Hattori, T., Terada, T. and Hamasuna, S. (1995) Regulation of the *Osem* gene by abscisic acid and the transcriptional activator VP1: analysis of *cis*-acting promoter elements required for regulation by abscisic acid and VP1. *The Plant Journal* 7, 913-925.

Heimovaara-Dijkstra, S., Mundy, J. and Wang, M. (1995) The effect of intracellular pH on the regulation of the RAB16A and the alpha-amylase 1/6-4 promoters by abscisic acid and gibberellia. *Plant Molecular Biology* 27, 815-820.

Hill, A., Nantel, A., Rock, C.D. and Quatrano, R.S. (1996) A conserved domain of the *viviparous-1* gene product enhances the DNA binding activity of the bZIP protein EmBP-1 and other transcription factors. *Journal of Biological Chemistry* 271, 3366-3374.

Hughes, D.W. and Galau, G.A. (1991) Developmental and environmental induction of Lea and LeaA mRNAs and the post-abscission program during embryo culture. *The Plant Cell* 3, 605-618.

Izawa, T., Foster, R. and Chua, N.-H. (1993) Plant bZIP protein DNA binding specificity. *Journal of Molecular Biology* 230, 1131-1144.

Jensen, A.B., Busk, P.K., Figueras, M., Mar Albà, M., Peracchia, G., Messeguer, R., Goday, A. and Pagès, M. (1996) Drought signal transduction in plants. *Plant Growth Regulation* 20, 105-110.

Jonak, C., Heberle-Bors, E. and Hirt, H. (1994) MAP kinases: universal multi-purpose signal tools. *Plant Molecular Biology* 24, 407-416.

King, R.W. (1976) Abscisic acid in developing wheat grains and its relationship to grain growth and maturation. *Planta* 132, 43-61.

Koornneef, M. (1986) Genetic aspects of abscisic acid. In: Blonstein A.D. and King P.J. (eds), *A Genetic Approach to Plant Biochemistry*. Springer Verlag, Vienna, pp. 35-54.

Koornneef, M., Hanhart, C.J., Hilhorst, H.W.M. and Karssen, C.M. (1989) *In vivo* inhibition of seed development and reserve protein accumulation in recombinants of abscisic acid biosynthesis and responsiveness mutants in *Arabidopsis thaliana*. *Plant Physiology* 90, 463-469.

Léon-Kloosterziel, K.M., van de Bunt, G.A., Zeevart, J.A.D. and Koornneef, M. (1996) *Arabidopsis* mutants with a reduced seed dormancy. *Plant Physiology* 110, 233-240.

Leone, A., Costa, A., Tucci, M. and Grillo, S. (1994) Comparative analysis of short- and long-term changes in gene expression caused by low-water potential (*Solanum tuberosum*) cell-suspension cultures. *Plant Physiology* 106, 703-712.

Marin, E., Nussaume, L., Quesada, A., Gonneau, M., Sotta, B., Hugueney, P., Frey, A. and Marion-Poll, A. (1996) Molecular identification of zeaxanthin epoxidase of *Nicotiana plumbaginifolia*, a gene involved in abscisic acid biosynthesis and corresponding to the *ABA* locus of *Arabidopsis thaliana*. *EMBO Journal* 15, 2331-2342.

McCarty, D.R. (1995) Genetic control and integration of maturation and germination pathways in seed development. *Annual Review of Plant Physiology and Plant Molecular Biology* 46, 71-93.

Mundy, J. and Chua, N.-H. (1988) Abscisic acid and water-stress induce the expression of a novel rice gene. *EMBO* 7, 2279-2286.

Nelson, D., Salamini, F. and Bartels, D. (1994) Abscisic acid promotes novel DNA-binding activity to a desiccation-related promoter of *Craterostigma plantagineum*. *The Plant Journal* 5, 451-458.

Niogret, M.F., Culiáñez-Macià, F.A., Goday, A., Albà, M.M. and Pagès, M. (1996) Expression and cellular localization of *rab28* mRNA and Rab28 protein during maize embryogenesis. *The Plant Journal* 9, 549-557.

Parcy, F., Valon, C., Raynal, M., Gaubier-Comella, P., Delseny, M. and Giraudat, J. (1994) Regulation of gene expression programs during *Arabidopsis* seed development: roles of the *ABI3* locus and of endogenous abscisic acid. *The Plant Cell* 6, 1567-1582.

Pla, M., Goday, A., Vilardell, J., Gómez, J. and Pagès, M. (1989) Differential regulation of ABA-induced 23-25 kDA proteins in embryo and vegetative tissues of the viviparous mutants of maize. *Plant Molecular Biology* 13, 385-394.

Pla, M., Gómez, J., Goday, A. and Pagès, M. (1991) Regulation of the abscisic acid-responsive gene *rab28* in maize *viviparous* mutants. *Molecular and General Genetics* 230, 394-400.

Pla, M., Vilardell, J., Guiltinan, M.J., Marcotte, W.R., Niogret, M.F., Quatrano, R.S. and Pagès, M. (1993) The *cis*-regulatory element CCACGTGG is involved in ABA and water-stress responses of the maize gene *rab28*. *Plant Molecular Biology* 21, 259-266

Robertson, D.S. (1955) The genetics of vivipary in maize. *Genetics* 40, 745-760.

Rock, C.D., Heath, T.G., Gage, D.A. and Zeevart, J.A.D. (1991) Abscisic alcohol is an intermediate in abscisic acid biosynthesis in a shunt pathway from abscisic aldehyde. *Plant Physiology* 97, 670-676.

Salinas, J., Oeda, K. and Chua, N.-H. (1992) Two G-box-related sequences confer different expression patterns in transgenic tobacco. *The Plant Cell* 4, 1485-1493.

Sheen, J. (1996) Ca^{2+}-dependent protein kinases and stress signal transduction in plants. *Science* 274, 1900-1902.

Shen, Q., Zhang, P. and Ho, T.-H.D. (1996) Modular nature of abscisic acid (ABA) response complexes: composite promoter units that are necessary and sufficient for ABA induction of gene expression in barley. *The Plant Cell* 8, 1107-1119.

Skriver, K. and Mundy, J. (1990) Gene expression in response to abscisic acid and osmotic stress. *The Plant Cell* 2, 503-512.

Stockinger, E.J., Gilmour, S.J. and Thomashow, M.F. (1997) *Arabidopsis thaliana CBF1* encodes an AP2 domain-containing transcriptional activator that binds to the C-repeat/DRE, a *cis*-acting DNA regulatory element that stimulates transcription in response to low temperature and water deficit. *Proceedings of the National Academy of Sciences USA* 94, 1035-1040.

Tardieu, F. (1996) Drought perception by plants. Do cells of droughted plants experience water stress? *Plant Growth Regulation* 20, 93-104.

Urao, T., Katagiri, T., Mizoguchi, T., Yamaguchi-Shinozaki, K. and Shinozaki, K. (1994) Two genes that encode Ca^{2+}-dependent protein kinases are induced by drought and high-salt stresses in *Arabidopsis thaliana*. *Molecular and General Genetics* 244, 331-340.

Vilardell, J., Goday, A., Freire, M.A., Torrent, M., Martinez, C., Torné, J.M. and Pagès, M. (1990) Gene sequence, developmental expression, and protein phosphorylation of RAB-17 in maize. *Plant Molecular Biology* 14, 423-432.

Vilardell, J., Martínez-Zapater, J.M., Goday, A., Arenas, C. and Pagès, M. (1994) Regulation of the rab17 gene promoter in transgenic *Arabidopsis* wild-type, ABA-deficient, and ABA-insensitive mutants. *Plant Molecular Biology* 24, 561-569.

Williamson, J.D. and Scandalios, J.G. (1992) Differential response of maize catalases to abscisic acid: Vp1 transcriptional activator is not required for abscisic acid-regulated *Cat1* expression. *Proceedings of the National Academy of Sciences of the USA* 89, 8842-8846.

Yamaguchi-Shinozaki, K. and Shinozaki, K. (1994) A novel *cis*-acting element in an *Arabidopsis* gene is involved in responsivenes to drought, low-temperature, or high-salt stress. *The Plant Cell* 6, 251-264.

Zeevaart, J.A.D. and Creelman, R.A. (1988) Metabolism and physiology of abscisic acid. *Annual Review of Plant Physiology and Plant Molecular Biology* 39, 439-473.

13 Manipulation of Growth of Horticultural Crops Under Environmental Stress

W.J. Davies, D.S. Thompson and J.E. Taylor

Department of Biological Sciences, I.E.N.S., Lancaster University, Bailrigg, Lancaster LA1 4YQ, UK

Introduction

Although growth is the most obvious of all plant activities and is essential for the production of economic yield, we still know relatively little about the processes limiting cell division and enlargement under environmental stress. Much of what we call plant growth is actually increased water content, and it is therefore not surprising that when water availability becomes limiting, as it does under many different kinds of environmental stress, growth is often restricted. It is not always clear, however, if this is because the swelling of the cell is limited by restricted water supply or because the properties of the cell walls have been changed by the stress, making them less able to extend.

In much published work, the former was commonly assumed to be the case, often without direct measurements of the water relationships of expanding cells. These measurements are not easy to make because growing regions are often of a restricted size but more so because growing cells can continue to grow during the period of measurement. If the plant tissue is sampled and removed from a supply of water, as it will be when placed in a psychrometer, then growing cells will lose turgor as the walls relax to the yield threshold. The water relationship value measured will reflect this state rather than the water status of the plant before it was sampled (Cosgrove, 1985). More recently, following the development of techniques that reliably can assess the water relationships of the growing regions of plants, we have discovered that the water status of these cells often is tightly regulated. During drought and indeed under other stresses, growth rate will decline even when the turgor of growing cells is entirely sustained (e.g. Michelena and Boyer, 1982).

During the last 15 years, a variety of techniques has been employed to break the link between soil drying and the restriction of water supply to the shoots. For example, one of these techniques involves pressurizing the roots of intact plants to counterbalance the declining water potential of the soil as it dries. This technique will sustain shoot water relationships of droughted plants in a state comparable with those of well-watered plants, but, despite this, growth and stomatal conductance of plants with roots in drying soil are still restricted (see, e.g., Passioura, 1988; Gollan *et al.*, 1992).

Results from the range of studies reported above and from other related studies have focused attention on the possibility that growth and functioning of shoots of plants in drying soil and under other stresses may be regulated by chemical influences that can move from the roots in the transpiration stream (see, e.g., Davies and Zhang, 1991; Davies *et al.*, 1994). The hypothesis is that the control of growth may reside in the chemical moderation of the properties of the cell walls of the plant. In this chapter, we highlight examples of chemical control of growth and functioning of plants under environmental stress and show how 'chemical signalling' may be manipulated by the grower for economic advantage. We also show how cell wall properties may be manipulated to modify the growth rate and the final yield of fruit crops.

Endogenous Chemical Regulators and the Manipulation of Growth Under Stress

A major problem with many fruit crops is excessive vigour as a result of continuous shoot growth. Use of assimilates in leaf growth restricts fruit set and development, while excessive leaf area can lead to wasteful water loss and the development of damaging water deficits. Too many leaves on plants in cooler, moister regions can predispose plants to fungal diseases such as mildew of the leaves or fruit rots. In addition, shading of fruits causes a reduction in fruit quality. With many crops, shoot development is restricted by pruning, but this can be an expensive process and can provide infection sites for diseases.

Reduced water supply (sometimes referred to as regulated deficit irrigation, RDI) can significantly limit leaf growth, but this treatment generally results in yield reductions. For example, deficit irrigation with grapes can result in decreases in pruning weight (leaf area development) of 15% or more, but these reductions are associated with yield decreases of the same order due to decreased berry weight (Matthews and Anderson, 1988, 1989; Goodwin and Jerie, 1992). This approach to saving water and reducing shoot growth can also result in splitting of fruit, or other types of fruit blemishing which can reduce the value of the crop. Furthermore, to operate RDI effectively, the grower is required to monitor plant and soil moisture status, and this can require sophisticated, expensive equipment. An irrigation system capable of delivering precisely monitored and directed quantities of water on demand is also required. This too can be expensive.

Recently, Gowing *et al.* (1990) have shown that production and expansion of leaves of apple trees can be restricted significantly by watering only half the root system of the plant. This restriction occurs without any obvious influence of the treatment on shoot water relationships. These authors were able to reverse the drought-induced limitation in shoot growth by cutting from the plant the roots that were in contact with the dry soil (Fig. 13.1). It seems very unlikely

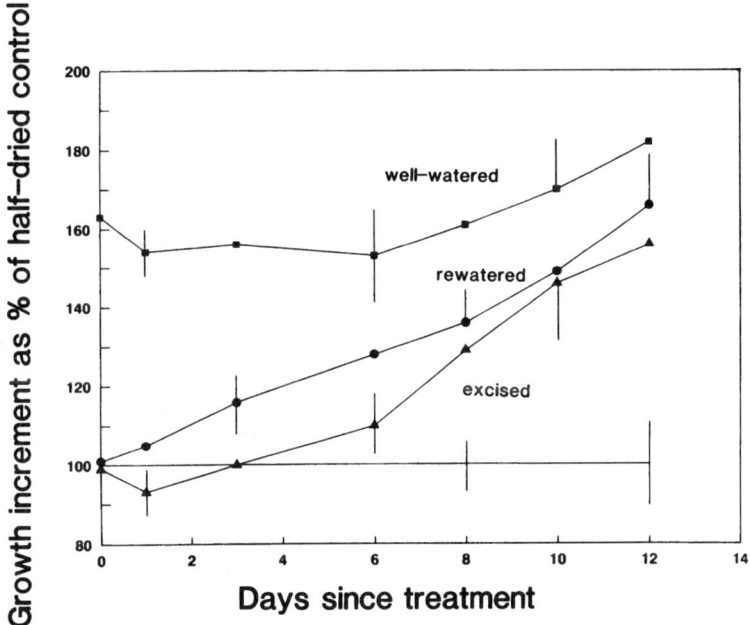

Fig. 13.1. Growth of apple shoots on plants with split roots, expressed as a percentage of growth increment shown by plants where half of the original root system has received no water for 23 days. Well-watered plants (●) were watered on both halves of the root system for 23 days prior to the period shown and for the whole of the experimental period. Re-watered plants (■) received no water on half of each root system for 23 days, after which both halves of each root system received water (day 0). Excised plants (▲) received no water on half of each root system for 23 days, after which roots in contact with dry soil were excised from the plants with a razor blade (day 0). Points are 3-day pooled means ($n = 6$). The vertical bar represents the LSD ($P < 0.05$) between the control and the root excised treatment. (From Gowing et al., 1990.)

that removal of roots could have made more water available to the shoots and so the interpretation of the result was that: (i) soil drying had generated an increased delivery of chemical regulators to the shoots in the transpiration stream, (ii) this chemical 'signal' had arisen as a result of interaction between the plant roots and the drying soil, (iii) the chemical signal resulted in restricted leaf growth even though the water relationships of the shoot were unperturbed by the soil drying treatment, and (iv) removal of the roots in contact with drying soil removed the source of the chemical regulator, reduced flux to the shoots and allowed the growth of the shoot to recover to the rate shown by the well-watered controls.

ABA signalling and growth

A significant amount of research over the last 10-12 years has been devoted to the identification and quantification of chemical signals moving from roots to shoots in plants in drying soil. Davies et al. (1994) have summarized recent progress in this field. In many droughted plants, an increased flux of the plant growth regulator abscisic acid (ABA) moves from roots to shoots and apparently regulates both stomatal behaviour (Zhang and Davies, 1991) and leaf growth (Zhang and Davies, 1990). The ABA signal may provide the shoots with a 'measure' of the access that the roots have to soil water (Tardieu et al., 1992) such that plants can modify their growth and development accordingly. There is some debate over the information content of the signal, i.e. exactly what is being 'measured', and there is also some controversy over the nature of the ABA signal. Many authors have shown good correlations between either growth or stomatal behaviour and the concentration of ABA in the xylem stream. It is easy to break these correlations, however, and this is because the xylem stream may be several compartments away from the site of action of the hormone. Hartung et al. (1998) have explained how the distribution of ABA in the leaf and, therefore, the accumulation at the site of action will be influenced by the transpiration flux, the rate of metabolism of the hormone, the pH gradients in the leaf and the activities of a range of hormone carriers. Other plant growth regulators (e.g. Fussader et al., 1992), modified ion fluxes (Gollan et al., 1992), electrical signals and even the water status of the leaf (Dodd and Davies, 1996) can also contribute to drought signalling. Given all of the factors contributing to chemical signalling of the effects of drought, it is not surprising that there has been much controversy over the adequacy or otherwise of particular signals (e.g. Munns et al., 1993), and this is an area of continuing research, with some emphasis on the possibility that much ABA may be transported through the plant in a conjugated form which can be broken down to release ABA in the leaf (e.g. Netting et al., 1992).

We have seen above that a variety of ingenious systems have been used to establish the existence of chemical regulation of shoot growth and functioning under drought. This work has also involved the use of genetic tools, particularly the use of low-ABA mutants (see, e.g., Taylor, 1991). Very recently, a cDNA clone encoding zeaxanthin epoxidase (ZE) was isolated and sequenced (Marin et al., 1996). This represented the first DNA sequence data on an ABA biosynthesis enzyme. This cDNA has been used to provide a heterologous probe to isolate a ZE cDNA from tomato (Burbidge et al., 1997a). Interestingly, Northern analysis has revealed no significant effect of a drought treatment on ZE mRNA. More recently, Burbidge et al. (1997b) have described the structure and expression of a cDNA encoding a putative neoxanthin cleavage enzyme (NCE). Water deficit dramatically increases the production of NCE mRNA, suggesting that this may be a key regulatory step in the biosynthesis of ABA. It is most likely that tools of this kind will be used increasingly to elucidate further the role of plant hormones in plant stress biology.

pH signalling

While much attention has been given to defining and quantifying the ABA signal, the pH of the xylem sap is another chemical signal which not only varies as a function of soil drying but also can change as other conditions around the roots change. In addition, this is a variable that can easily be manipulated artificially, for example in tomatoes growing in rockwool or flowing solution culture. Schurr *et al.* (1992) have reported on drought-induced alkalinization of the xylem sap of sunflower, and Hartung *et al.* (1998) have shown how this will be expected to alter the delivery of ABA to sites of action on the guard cells. Wilkinson and Davies (1997) have fed artificial xylem sap buffered to different pHs to detached leaves and shown how stomatal closure induced by high pH requires the presence of a low ABA concentration which does not itself influence the stomata. High xylem sap pH slows the loading of ABA from the apoplast to the symplast, resulting in relatively high apoplastic ABA concentrations close to a site of action on the guard cell. Thompson *et al.* (1997) have suggested that drought-induced high pH signals may also limit leaf growth through an effect on the distribution of ABA in the growing leaf. In addition, the activity of wall-loosening enzymes may also be pH sensitive (see below) and, therefore, pH changes in the xylem sap may also have a direct effect on leaf growth.

Partial root drying and the regulation of growth

In 1991, Loveys suggested that it might be possible to exploit the split root system described above to regulate vegetative growth of vines. The hope was that root-sourced chemical regulators would restrict shoot development while the maintenance of shoot water status would ensure 'normal' fruit development. Subsequently, Dry *et al.* (1995) described results from experiments where grape vine vigour was substantially reduced without any yield penalty, and at the same time, efficiency of water use and quality of fruit were significantly improved (see also Fuller, 1997) (Table 13.1). These results were achieved by employing partial

Table 13.1. Effect of partial root zone drying on yield, water use, growth and canopy development of 'Cabernet Sauvignon' grapes grafted on to 'Ramsey'.

Parameter	Control	Treated	% Difference	Significance
Irrigation (ml ha^{-1})	0.92	0.56	−39	
Fruit weight (kg/vine)	4.73	4.88	+3	ns
Water use efficiency (g fruit l^{-1} irrig.)	4.9	7.2	+46	< 0.01
Pruning weight (kg/vine)	4.62	3.47	−25	< 0.01
Lateral leaf area (m^2/vine, harvest)	9.2	5.5	−41	< 0.05
Stomatal conductance (mmol m^{-2} s^{-1})	250	180	−29	< 0.05
Bunch exposure index (veraison)	120	325	+171	< 0.01
Colour (mg g^{-1} weight)	1.19	1.72	+45	< 0.05
Glycosyl-glucose (mol g^{-1} fresh weight)	2.64	3.75	+42	< 0.05

Experiments were performed at the Waite Campus of the University of Adelaide, South Australia during 1994/1995 (from Dry *et al.*, 1995).

root-zone drying (PRD) in different experiments over several seasons. Here, one side only of each row of vines was irrigated for 10 days. After this time, the irrigation was changed to the other side of the root system. The hypothesis is that in this way, the supply of growth regulators to the shoots is sustained, thereby resulting in a significant modification of shoot growth. Presumably, increased penetration of solar radiation to the fruit results in increased quality and value of the crop. In addition, despite the limitations in gas exchange imposed by the treatment, it seems likely that carbohydrates and other compounds are diverted from leaves to the fruit, thereby enhancing the accumulation of secondary metabolites (anthocyanins, phenolics, glycosylglucose) which contribute to flavour and aroma. These changes will also contribute to enhanced crop value. It is interesting to speculate on why the growth of the fruit is not affected by PRD. It may be because the chemical regulators in the xylem stream are not accumulated in the fruit, perhaps because of limited xylem transport to the fruit. Alternatively, the properties of the cell walls of the fruit and the leaves may be modified differently by chemical signals. This provides us with a justification for investigating cell wall properties of growing organs.

This important biochemical and molecular work will provide much greater insight into the regulation of fruit growth under stress if it is conducted in conjunction with a study of water relationships of the growing fruit. There are few data of this kind in the literature, partly because sophisticated technology is required to make significant advances. We have used the micropressure probe to measure directly the turgor of the fruit pericarp as the tomato plant is slowly droughted. Fruit growth is also measured continuously using an LVDT (Fig. 13.2). Growth rate declines slowly over 3 days as soil water supply is limited but, interestingly, fruit turgor remains largely unaffected over this period. This is in contrast to the turgor pressure of the leaf which declines steadily as the soil dries. The single-cell sampling facility on the probe allows for the collection of cell sap. Picolitre osmometry was used to measure the osmotic pressure of this sap so that cell water potential could be calculated. Taken together, these data suggest that the tomato fruit is relatively well buffered against the development of cell water deficit as the soil dries. It is necessary for us to explain the reduction in growth which occurs despite the effective maintenance of pericarp turgor. These data provide extra justification for investigating the effects of soil drying on the properties of cell walls in the fruit. The reduction in cell water potential and turgor in the leaves in response to soil drying raises the possibility that increased amounts of chemical regulators such as ABA move from leaves to limit fruit growth under drought. To sustain this hypothesis, it will be necessary to quantify the flux of growth regulators from leaves to fruit in both well-watered and droughted conditions.

Cellular Properties and the Growth of Leaves and Fruit

Whether growth of leaves and fruit is responding to chemical or hydraulic signalling, there is general agreement in the literature that much growth

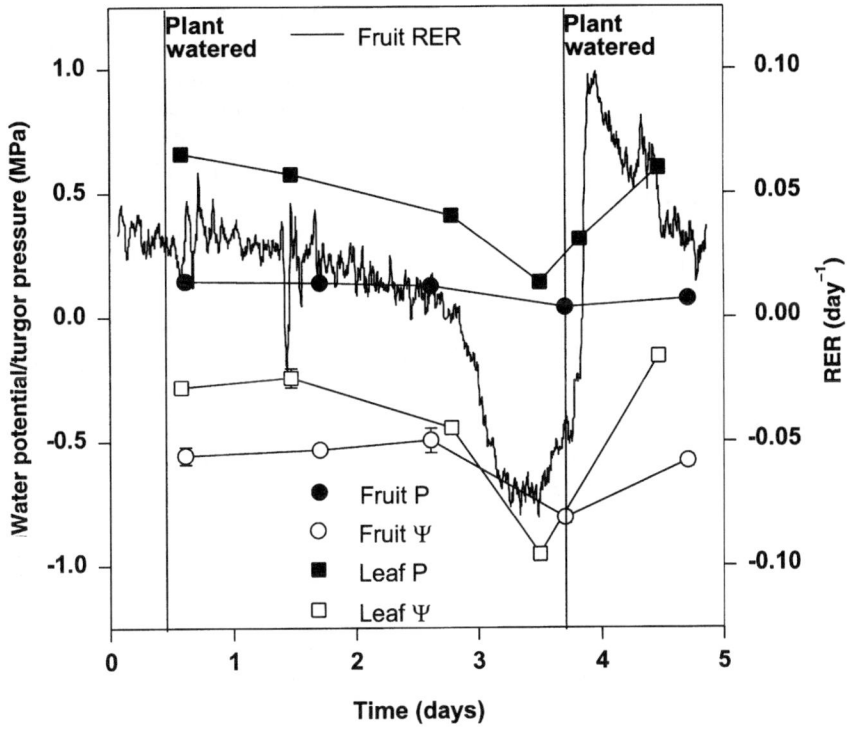

Fig. 13.2. Effects of soil drying on the growth rate and water relationships of tomato fruit. The plant was last watered on day 0, and the soil was dried gradually until late on day 3, at which point the plant was re-watered. Fruit relative growth rate (RER) was measured with an LVDT while fruit pericarp and leaf turgor were measured with the micropressure probe. Water potentials were calculated by difference from cell turgors and osmotic pressures measured by single-cell sampling.

regulation occurs as a result of modification of the properties of the cell walls (e.g. McDonald and Davies, 1996). One consideration before attempting to quantify cell wall properties is to decide which tissues are likely to be limiting organ growth. In many plant organs, the inner tissue expands once the epidermis is peeled (e.g. Kutschera, 1989), and this observation suggests that the cortical tissue is constrained by the epidermis. Pressure of the cortical cells on the epidermis has been called 'tissue pressure' (TP; Kutschera and Kohler, 1992) and it has been assumed that in expanding organs showing TP, the epidermis limits growth and the rate of organ growth is determined by the rate at which the epidermis releases TP (e.g. Kutschera, 1995). Our recent work (Thompson *et al.*, 1998) with tomato fruit suggests that this model applies to expanding tomato fruit and that the rate of growth and the final fruit size are determined by a combination of the water relationships of the pericarp as they contribute to the TP and the cell wall properties of the epidermis.

In considering the wall properties of the epidermal cells, we have been concerned both with wall-loosening and wall-stiffening enzymes (see, e.g., Thompson et al., 1997). In the literature, there is interest in the roles played by two putative wall-loosening enzymes, xyloglucan endoglycosylase (XET) (e.g. Fry et al., 1992) and a class of enzymes termed the expansins (see McQueen-Mason, 1995). Cell walls may be stiffened through an up-regulation of peroxidase enzymes (MacAdam et al., 1992).

XET and the regulation of growth

It is proposed that XET allows cell expansion by cleaving xyloglucan molecules that tether adjacent cellulose microfibrils in the cell wall. It may also rejoin the cut ends of cleaved tethers (Smith and Fry, 1989). The net effect of these processes may be the separation of cellulose microfibrils allowing controlled cell expansion. Several attempts have been made to correlate the spatial distribution of growth in leaves and roots with the spatial distribution of XET activity (e.g. Pritchard et al., 1993). As a result of these investigations, it is suggested that a peak in XET activity may be important in allowing a peak in growth rate but, in maize leaves, the relationship between the two variables is not straightforward, with maximal XET activity preceding maximal growth rate by about 50 h (Palmer and Davies, 1996). In roots of maize, the temporal separation of the peaks in activity is only 4 h (Wu et al., 1994). In leaves, XET activity is reduced as leaves age and their elongation rate falls with the reduction in enzyme activity preceding the fall in growth. However, there is a residual peak in XET activity after the cessation of leaf growth which argues against a simple causal relationship between the two variables. Nevertheless, it is of interest that soil-drying treatments do influence XET activity in both leaves and roots. In both tissues, mild soil drying leads to an up-regulation of enzyme activity close to the meristem. It is argued that this up-regulation leads to a maintenance of growth in this region (Wu et al., 1994; Thompson et al., 1997). Interestingly, this increased activity may be associated with drought-induced accumulation of ABA in this region.

There is some question of whether XET can really be considered to be a wall-loosening enzyme which regulates growth, since McQueen-Mason et al. (1993) found that this enzyme was unable to catalyse the expansion of boiled cell walls in an *in vitro* assay. One interpretation of all of these results is that XET activity may be required for growth, perhaps regulating the distribution of cell wall stress between xyloglucan chains cross-linking cellulose microfibrils, but that XET generally will not determine the rate of expansion. Results of our study of growing tomato fruit (Thompson et al., 1998) are consistent with this conclusion but do provide evidence of a rather more direct role for XET in the determination of growth rate. We assayed XET activity in the epidermis and in the pericarp. High activities declined as growth rate slowed (Fig. 13.3). Epidermal XET activity appeared to be proportional to the fruit relative growth rate (RGR) (Fig. 13.4), suggesting that under some circumstances the rate of fruit growth might be determined by the epidermal XET activity. This is in clear contrast to the situation found in leaves and roots. As in other plant organs, XET

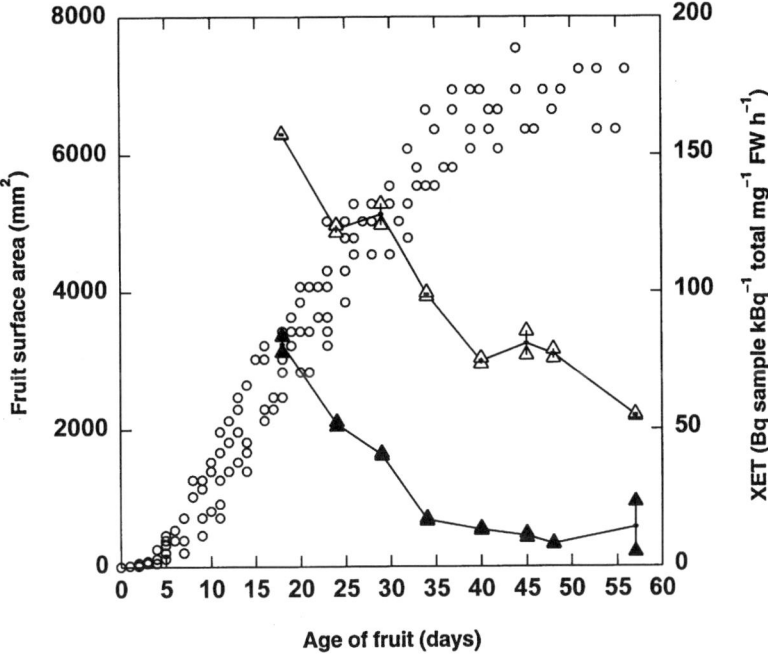

Fig. 13.3. XET activities extracted from the epidermis (△) and pericarp (▲) of tomato fruit at different stages of development. Duplicate measurements are shown for each age of fruit. The lines connect the mean values. The cumulative fruit surface areas of the eight fruits harvested for XET and peroxidase assay are also presented (○). (From Thompson et al., 1998.)

activity was present after tomato fruit expansion had ceased, suggesting a mechanism of growth termination other than a reduction in the activity of this enzyme.

Peroxidase and the regulation of growth

One mechanism of growth termination is cell wall peroxidation, with some peroxidases cross-linking wall polymers to inhibit growth. A number of authors have noted the appearance or increase of peroxidase isoenzymes as the rate of plant growth slows or halts (Goldberg et al., 1986; Schnabelrauch et al., 1996). We have detected peroxidase activity in epidermal extracts of tomato fruit but, significantly, not in the pericarp and not in the epidermis until between 40 and 45 days after anthesis (Thompson et al., 1998). The increase in activity was coincident with the cessation of tomato fruit growth (Fig. 13.5). It seems possible that the duration of expansion of the fruit may then be determined by the time at which the peroxidase activity appears. This, in combination with the initial cell number and the rate of growth, may then combine to determine the final size of the fruit, an important variable for the grower. Premature up-

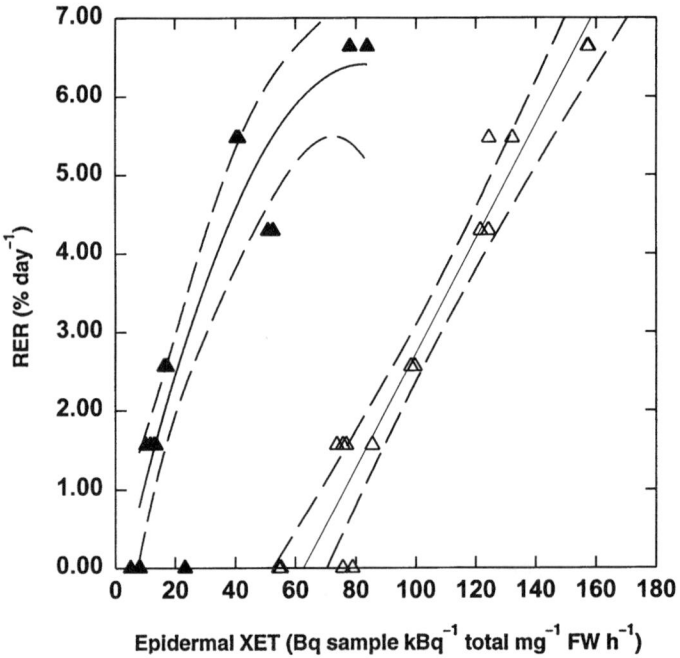

Fig. 13.4. The relationship between fruit surface area RER (relative expansion rate) and XET activities of the epidermis (△) and pericarp (▲) of tomato fruit at different stages of development. Duplicate measurements are shown. The unbroken lines are curves fitted to the epidermal and pericarp data by Sigmaplot by minimizing the squares of the residuals. The curve fitted to the epidermal data is first order and that fitted to the pericarp data is second order. The activities of the fruit harvested 57 days after anthesis were not fitted. The broken lines are 95% confidence limits. (From Thompson et al., 1998.)

regulation of peroxidase activity, perhaps by high temperature, a stress known to restrict the final size of tomato fruit, may be an early limitation on the capacity of the epidermis to expand. This premature stiffening combined with high tissue pressure could result in splitting of fruit, which obviously will reduce the value of the crop.

Bacon *et al.* (1997) have investigated the effects of drought on the spatial distribution of growth and cell wall peroxidase activity in the elongation zone of *Lolium* leaves. Abrupt increases in peroxidase activity occurred at the position where elongation ceased in the leaf meristems of well-watered and mildly droughted plants. More prolonged drought caused a 200–300% increase in cell wall-associated peroxidase activity in the elongation zone only. In another study, Thompson *et al.* (1997) were able to enhance cell wall-associated peroxidase activity in the growing zone of barley leaves by feeding elevated concentrations of ABA to barley leaves in a leaf growth bioassay. It seems clear then from studies

Growth of Crops Under Environmental Stress 167

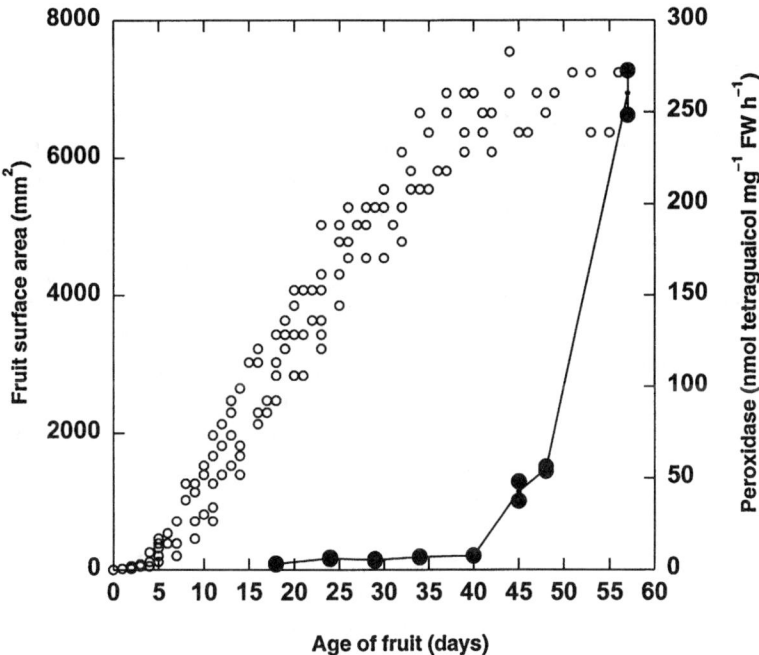

Fig.13.5. Weakly ionically bound peroxidase activities extracted from the epidermis of tomato fruit at different stages of development (●). Duplicate measurements are shown for each age of fruit. The lines connect the mean values. The cumulative fruit surface areas of eight fruits harvested for XET and peroxidase activity are also presented (○). (From Thompson *et al.*, 1998.)

of XET and peroxidase activities in droughted plants that chemical signalling can directly modify the properties of cell walls via a moderation of the activity of cell wall-associated enzymes.

Expansins and growth

Another possibility for chemical regulation of growth of droughted plants is the moderation of activity of the expansins. These enzymes appear to disrupt hydrogen bonds to cellulose microfibrils, allowing cell walls to relax, but they require acid conditions to do so. We have noted above how soil drying can lead to an alkalinization of the xylem sap. The cell walls of the leaf may also experience modified pH when soil and roots are dried. Van Volkenburgh and Boyer (1985) have used a flat-tipped pH electrode to measure the pH of an abraded surface of maize leaves and have detected significant alkalinization at low water potentials. If expansin action is required for growth, then an increase in cell wall pH under drought may be a primary mechanism restricting cell expansion.

Manipulation of Cell Wall Properties

It is clear from the preceding discussion that there are many situations in which an ability to manipulate cell wall growth could be of benefit to the grower and ultimately to the consumer. It may be possible to alleviate a restriction in growth or intervene to sustain growth under environmental stress. Sites of interest must include modification of the signalling processes used by the plant in response to water stress, and the downstream targets of such signalling pathways such as the enzymes shown to have a role to play in mediating cell growth. In the last few years, we have achieved a greatly increased understanding of both intercellular and intracellular signalling processes, but in most situations we still lack the degree of knowledge necessary directly and specifically to manipulate signalling pathways. More feasible sites for manipulation are the enzymes that show a correlative change in activity associated with a change in growth rate. Identification of clones encoding XET (Arrowsmith and de Silva, 1995), expansins (Shcherban et al., 1995) and peroxidase (Lagrimini et al., 1987) means that molecular techniques are possible routes for manipulating enzyme activity. Indeed a correlative change in enzyme activity does not prove a causal link with growth, whereas the spatial and/or temporal manipulation of growth and the expression of the genes encoding key enzymes can lend considerable weight to the proposal for functionality of the enzyme.

XET

A number of clones encoding XETs have now been isolated (de Silva et al., 1993; Saab and Sachs, 1995; Arrowsmith and de Silva, 1995; Rose et al., 1996). Le EXT, a clone encoding an XET from tomato, was found to be present primarily in elongating regions of the hypocotyl, and expression was induced by hormone treatments that induced extension growth in this tissue. The time scale of induction suggested that induction of this gene was not required for rapid growth responses but may be involved in sustained cell growth (Catala et al., 1997). In situ hybridization studies have indicated that expression of this clone is localized to the outer cell layers of the hypocotyl within the elongation zone. In nasturtium, there appear to be at least two different genes coding for XET, and these have very different patterns of expression; NGX1 (de Silva et al., 1993) is expressed in germinating seed cotyledons and XET1 (Rose et al., 1996) is expressed in all vegetative tissues except the cotyledons. These studies indicate that the different forms of XET may have different roles within the plant, one of which may be the regulation of cell extension.

Expansins

Shieh et al. (1997) used gene-specific probes to examine the regulation of expression of two transcripts encoding expansin. Their results suggest that changes in expansin expression are involved in some but not all growth responses in cucumber hypocotyls. Four rice expansin cDNAs have now been identified (Cho and Kende, 1997) and have been shown to have well-defined spatial expression patterns. Expression of the rice clones has been shown to be influenced by stimuli that also induce cell elongation, and these expansins are thought to be involved in the extension growth seen in rice in response to submergence. Similarly, Shimizu et al (1997) have been able to demonstrate using a reverse transcriptase-polymerase chain reaction (RT-PCR) approach that the expression of an mRNA for expansin in cotton fibre cells is high during cell elongation but decreases when cell elongation ceases.

One approach often adopted when trying to determine how genes are regulated is to look for functional motifs within the sequence. The expansin sequences examined so far appear to code for novel proteins in that they are without previously known functional motifs (Cosgrove, 1996). These studies have now identified a number of different expansin clones which appear to be more highly expressed in actively elongating tissues. Future studies may give more insight into how these different clones are temporally and spatially expressed and the factors that determine their functional role. As yet, we have little information as to how environmental stress might influence their expression.

Peroxidases

There are many different isoenzymes of peroxidase within individual plant species, which may have a role to play in lignification, disease resistance and the introduction of cross-links within cell walls. A number of peroxidase clones have now been identified (Lagrimini et al., 1987; Botella et al., 1994; Teichmann et al., 1997), and some of these have provided constructs that have been used to make antisense and overexpressing transgenic plants. Antisense tobacco plants deficient in an anionic peroxidase (Lagrimini et al., 1997a) have been show to display altered growth characteristics. Antisense expression was shown to lead to a significant decrease in peroxidase activity of up to 1600-fold. These plants had an apparently normal morphology but, from measurements taken of plant height 12 weeks after sowing, the antisense plants were significantly taller than the non-transformed controls. This increase in growth was also apparent as an increase in leaf thickness in antisense plants. In particular, measurements of peroxidase activity in antisense plants compared with the non-transformed controls demonstrated that the greatest reduction in enzyme activity was in the leaf epidermis. These changes in peroxidase activity were without effect on the degree of lignification of tissues within the transformed plants, or upon the isoenzymes that were induced in response to wounding. This demonstration of a localization in the down-regulation of peroxidase activity that enables plant to increase their growth rate suggests that it should be feasible to apply this type of approach to horticultural crops in which peroxidase activity is thought to limit cell and hence plant growth.

This peroxidase enzyme is not normally found in the roots of tobacco plants. Overexpression, in contrast to down-regulation, of the anionic peroxidase in tobacco produced plants which had a normal appearance above ground, but which wilted on reaching maturity (Lagrimini et al., 1997b). Wilting occurred even though the plants were shown to have functional stomata and a normal vascular anatomy and physiology. The transgenics, however, differed from wild-type plants in having considerably less root mass which was the result of an inhibition of root branching. The authors suggested that such an effect may be the result of an alteration in auxin metabolism and hence changes in the auxin to cytokinin ratio within the root tissue, leading to a reduction in root branching. A number of cDNAs coding for anionic peroxidases in other species such as tomato (Botella *et al.*, 1994) and maize (Teichman *et al.*, 1997) have been shown by Northern analysis to be located exclusively in the roots and expressed specifically in the root epidermis and subepidermal layers. The suggestion from this work is that these anionic peroxidases may be involved in tuberization. Manipulation of such enzymes may have an important impact on the ability of a plant to absorb water.

Conclusions

All of the molecular studies of enzymes thought to be involved in the regulation of extension growth have highlighted two features: firstly that each of these enzymes is encoded by more than one gene, each of which appears to have specific spatial regulation of expression; and secondly that expression of a number of those genes is localized to the tissues which our data suggest is of importance in terms of the control of growth, namely the epidermis. Although at this time little is known of how patterns of expression change in response to stress, the transgenic experiments of Lagrimini's group demonstrate that the genetic manipulation of peroxidase at least can lead to a change in the pattern of cell growth, and lend support to the idea that biotechnological approaches may in the future allow us to gain a better understanding of how plants regulate their growth. The molecular studies cited here also indicate how important it is to identify which gene, from each of these small gene families, is the key controlling element for growth in the particular tissue under study.

Although transgenic manipulations may be an effective way to increase the quality and yield of horticultural crops, consumer pressure may limit the introduction of such technology. Nevertheless, manipulations of the type described above will provide great insight into the processes regulating growth and it may be impossible to quantify the importance of particular enzymes without these tools. Results from these studies will also allow simple manipulations of growth conditions to be focused at key regulatory features. The commercial use of the split root technology to increase the profitability of the wine grape industry illustrates this point well. Advances in our understanding of the mechanisms used by plants to detect and respond to soil drying were necessary to allow the design of a simple cultural manipulation to modify growth. It is most unlikely that this manipulation would have been tried without our recognition of the importance of root to shoot communication in the plant's drought response. The success of this growing system shows how important it can be to fund basic science in an applied context.

References

Arrowsmith, D.A. and de Silva, J. (1995) Characterization of two tomato fruit-expressed cDNAs encoding xyloglucan *endo*-transglycosylase. *Plant Molecular Biology* 28, 391-403.

Bacon, M.A., Thompson, D.S. and Davies W.J. (1997) Can cell wall peroxidase activity explain the leaf growth response of *Lolium temulentum* L. during drought? *Journal of Experimental Botany* 48, 2075-2085.

Botella, M.A., Quesada, M.A., Kononowicz, A.K., Bressan, R.A., Pliego, F., Hasegawa, P.M. and Valpuesta, V. (1994) Characterisation and *in situ* localisation of a salt-induced tomato peroxidase mRNA. *Plant Molecular Biology* 25, 105-114.

Burbidge, A., Grieve, T.M., Terry, C., Corlett, J., Thompson, A., and Taylor, I.B. (1997a) Structure and expression of a cDNA encoding zeaxanthin epoxidase isolated from a wilt-related tomato (*Lycopersicon esculentum* Mill.) library. *Journal of Experimental Botany* 48, 1749-1750.

Burbidge, A., Grieve, T.M., Jackson, A., Thompson, A. and Taylor, I.B. (1997b) Structure and expression of a cDNA encoding a putative neoxanthin cleavage enzyme (NCE), isolated from a wilt-related tomato (*Lycopersicon esculentum* Mill.) library. *Journal of Experimental Botany* 47, 2111-2112.

Catala, C., Rose, J.K.C. and Bennett, A.B. (1997) Auxin regulation and spatial localisation of an endo-1,4-ß-D-glucanase and a xyloglucan endotransglycosylase in expanding tomato hypocotyls. *The Plant Journal* 12, 417-426.

Cho, H-T. and Kende, H. (1997) Expression of expansin genes in rice is correlated with growth. *Plant Physiology* 114 (Suppl.), 810.

Cosgrove, D.J. (1985) Cell wall yielding properties of growing tissues. Evaluation by *in vivo* stress relaxation. *Plant Physiology* 78, 347-356.

Cosgrove, D.J. (1996) Plant-cell enlargement and the action of expansins. *BioEssays* 18, 533-540.

Davies, W.J. and Zhang, J. (1991) Root signals and the regulation of growth and development of plants in drying soil. *Annual Review of Plant Physiology and Plant Molecular Biology* 42, 55-76.

Davies, W.J., Tardieu, F. and Trejo, C.L. (1994) How do chemical signals work in plants that grow in drying soil? *Plant Physiology* 104, 309-314.

de Silva, J., Jarman, S., Arrowsmith, D.A., Stronach, M.S., Chengappa, S., Sidebottom, C. and Reid, J.S.G. (1993) Molecular characterisation of a xyloglucan-specific endo-(1-4)-ß-D-glucanase (xyloglucan endotransglycosylase) from nasturtium seeds. *The Plant Journal* 3, 701-711.

Dodd, I.C. and Davies, W.J. (1996) The relationship between leaf growth and ABA accumulation in the grass leaf elongation zone. *Plant, Cell and Environment* 19, 1047-1056.

Dry, P., Loveys, B.R., Botting, D. and During, H. (1995) Effects of partial root drying on grapevine vigour, yield, composition of fruit and use of water. *Proceedings of the Australian Wine Industry Technical Conference* 9, 128-131.

Fry, S.C., Smith, R.C., Renwick, K.F., Martin, D.J., Hodge, S.K. and Matthews, K.J. (1992) Xyloglucanendotransglycosylase, a new wall-loosening enzyme activity from plants. *Biochemical Journal* 282, 821-828.

Fuller, P. (1997) Less water, more grapes, better quality: an ecological breakthrough in viticultural science. *Wine Industry Journal* 12, 155-157.

Fussader, A., Wartinger, A., Hartung, W., Schulze, E.-D. and Heilmeier, H. (1992) Cytokinins in the xylem sap of desert grown almond (*Prunus dulcis*) trees: daily courses and their possible interactions with abscisic acid and leaf conductance. *New Phytologist* 122, 45-52.

Goldberg, R., Imberty, A. and Chu-Ba, J. (1986) Development of isoperoxidases along the growth gradient in the mung bean hypocotyl. *Phytochemistry* 25, 1271-1274.

Gollan, T., Schurr, U. and Schulze, E.-D. (1992) Stomatal responses to drying soil in relation to changes in xylem sap composition of *Helianthus annuus* L. The concentration of cations, anions and amino acids and the pH of the xylem sap. *Plant, Cell and Environment* 15, 551-559.

Goodwin, I. and Jerie, P. (1992) Regulated deficit irrigation: from concept to practice. Advances in vineyard irrigation. *Australian and New Zealand Wine Industry Journal* 9, 258-261.

Gowing, D.J.C., Davies, W.J. and Jones, H.G. (1990) A positive root-sourced signal as an indicator of soil drying in apple *Malus* × *domestica* Borkh. *Journal of Experimental Botany* 41, 1535-1540.

Hartung, W., Radin, J.W. and Hendrix, D.L. (1988) Abscisic acid movement into the apoplastic solution of water stressed cotton leaves. *Plant Physiology* 86, 908-913.

Hartung, W., Wilkinson, S. and Davies, W.J. (1998) Factors that regulate abscisic acid concentration at the primary site of action on the guard cell. *Journal of Experimental Botany* 49, 361-367.

Kutschera, U. (1989) Tissue stresses in growing organs. *Physiologia Plantarum* 77, 157-163.

Kutschera, U. (1995) Tissue pressure and cell turgor in axial plant organs: implications for the organismal theory of multicellularity. *Journal of Plant Physiology* 146, 126-132.

Kutschera, U. and Kohler, K. (1992) Turgor and longitudinal tissue pressure in hypocotyls of *Helianthus annuus* L. *Journal of Experimental Botany* 43, 1577-1581.

Lagrimini, L.M., Burkhart, W.A., Moyer, M.B. and Rothstein, S. (1987) Molecular cloning of complementary DNA encoding the lignin-forming peroxidase from tobacco: molecular analysis and tissue specific expression. *Proceedings of the National Academy of Sciences USA* 84, 7542-7546.

Lagrimini, L.M., Gingas, V., Finger, F., Rothstein, S. and Lui, T.-T.Y. (1997a) Characterisation of antisense transformed plants deficient in tobacco anionic peroxidase. *Plant Physiology* 114, 1187-1196.

Lagrimini, L.M., Joly, R.J., Dunlap, J.R. and Lui, T.-T.Y. (1997b) The consequence of peroxidase overexpression in transgenic plants on root growth and development. *Plant Molecular Biology* 33, 887-895.

Loveys, B.R. (1991) How useful is a knowledge of ABA physiology for crop improvement? In: Davies, W.J. and Jones, H.G. (eds), *Abscisic Acid*. BIOS Scientific Publishers, Oxford, pp. 245-260.

MacAdam, J.W., Nelson, C.J. and Sharp, R.E. (1992) Peroxidase activity in the leaf elongation zone of tall fescue. 1. Spatial distribution of ionically bound peroxidase activity in genotypes differing in length of the elongation zone. *Plant Physiology* 99, 872-878.

Marin, E., Nussaume, L., Quesada, A., Gonneau, M., Sotta, B., Hugueney, P., Fry, A. and Marion-Poll, A. (1996) Molecular identification of zeaxanthin epoxidase of *Nicotiana plumbaginifolia*, a gene involved in abscisic acid biosynthesis and corresponding to the ABA locus of *Arabidopsis thaliana*. *EMBO Journal* 15, 2331-2342.

Matthews, M.A. and Anderson, M.M. (1988) Fruit ripening in *Vitis vinifera* L.: responses to seasonal water deficits. *American Journal of Enology and Viticulture* 39, 313-320.

Matthews, M.A. and Anderson, M.M. (1989) Reproductive development in grape (*Vitis vinifera* L.): responses to seasonal water deficits. *American Journal of Enology and Viticulture* 40, 52-60.

McDonald, A.J.S. and Davies, W.J. (1996) Keeping in touch: responses of the whole plant to deficits in water and nitrogen supply. *Advances in Botanical Research* 22, 229-300.

McQueen-Mason, S. (1995) Expansins and cell wall expansion. *Journal of Experimental Botany* 46, 1639-1650

McQueen-Mason, S., Fry, S.C., Durachko, D.M. and Cosgrove, D.J. (1993) The relationship between xyloglucan endotransglycosylase and *in vitro* cell wall extension in cucumber hypocotyls. *Planta* 190, 327-331.

Michelena, V.A. and Boyer, J.S. (1982) Complete turgor maintenance and low water potentials in the elongation regions of maize leaves. *Plant Physiology* 69, 1145-1149.

Munns, R., Passioura, J.B., Milborrow, B.V., James, R.A. and Close, T.J. (1993) Stored xylem sap from wheat and barley in drying soil contains a transpiration inhibitor with a large molecular size. *Plant, Cell and Environment* 16, 867-872.

Netting, A.G., Willows, R.D. and Milborrow, B.V. (1992) The isolation and identification of the prosthetic group released from a bound form of abscisic acid. *Plant Growth Regulation* 11, 327-334.

Palmer, S.J. and Davies, W.J. (1996) An analysis of the relative elemental growth rate, epidermal cell size and xyloglucan endotransglycosylase activity through the growing zone of aging maize leaves. *Journal of Experimental Botany* 47, 339-347.

Passioura, J.B. (1988) Root signals control leaf expansion in wheat seedlings growing in drying soil. *Australian Journal of Plant Physiology* 15, 687-693.

Pritchard, J., Hetherington, P.R., Fry, S.C. and Tomos, A.D. (1993) Xyloglucan endotransglycosylase activity, microfibril orientation and the profiles of cell wall properties along the growing regions of maize roots. *Journal of Experimental Botany* 44, 1281-1289.

Rose, J.K.C., Brummell, D.A. and Bennett, A.B. (1996) Two divergent xylogucan endotransglycosylases exhibit mutually exclusive patterns of expression in nasturtium. *Plant Physiology* 110, 493-499.

Saab, I.N. and Sachs, M.M. (1995) Complete cDNA and genomic sequence encoding a flooding-responsive gene from maize (*Zea mays* L.) homologous to xyloglucan *endo*transglycosylase. *Plant Physiology* 108, 439-440.

Schnabelrauch, L.S., Kieliszewski, M., Upham, B.L., Alizedeh, H. and Lamport, D. (1996) Isolation of pI 4.6 extensin peroxidase from tomato cell suspension cultures and identification of Val-Tyr-Lys as putative intermolecular cross link sites. *The Plant Journal* 9, 477-489.

Schurr, U., Gollan, T. and Sculze, E.-D. (1992) Stomatal responses to drying soil in relation to the changes in the xylem sap composition of *Helianthus annuus* II. Stomatal sensitivity to abscisic acid imported from the xylem sap. *Plant, Cell and Environment* 15, 561-567.

Shcherban, T.Y., Shi, J., Durachko, D.M., Guiltinan, M.J., McQueen-Mason, S.J., Shieh, M. and Cosgrove, D.J. (1995) Molecular-cloning and sequence analysis of expansins – a highly conserved multigene family of proteins that mediate cell wall extensin in plants. *Proceedings of the National Academy of Sciences USA* 92, 9245-9249.

Shieh, M.W., Shi, J. and Cosgrove, D.J. (1997) Developmental, hormonal and light regulation of the transcript for the cell wall loosening protein expansin. *Plant Physiology* 114 (Suppl.), 341.

Shimizu, Y., Aotsuksa, S., Hasegawa, O., Kawada, T. Sakuno, T., Sakai, F. and Hayashi, T.

(1997) Changes in levels of mRNAs for cell wall-related enzymes in growing cotton fiber cells. *Plant and Cell Physiology* 38, 375-378.

Smith, R.C. and Fry, S.C. (1989) Extracellular transglycosylation involving xyloglucan oligosaccharides *in vivo*. In: Fry, S.C. (ed.), *Fifth Cell Wall Meeting Book of Abstracts*. Scottish Cell Wall Group, Edinburgh.

Tardieu, F., Zhang, J. and Davies, W.J. (1992) What information is conveyed by an ABA signal from maize roots in drying field soil? *Plant, Cell and Environment* 15, 185-191.

Taylor, I.B. (1991) Genetics of ABA synthesis. In: Davies, W.J. and Jones, H.G. (eds), *Abscisic Acid*. BIOS Scientific Publishers, Oxford, pp. 23-37.

Teichmann, T., Guan, C., Kristoffersen, P., Muster, G., Tietz, O. and Palme, K. (1997) Cloning and biochemical characterisation of an anionic peroxidase from *Zea mays*. *European Journal of Biochemistry* 247, 826-832.

Thompson, D.S., Wilkinson, S., Bacon, M.A. and Davies W.J. (1997) Multiple signals and mechanisms that regulate leaf growth and stomatal behaviour during water deficit. *Physiologia Plantarum* 100, 303-313.

Thompson, D.S., Davies, W.J. and Ho, L.C. (1998) Is tomato fruit growth regulated by the epidermis? *Plant, Cell and Environment,* in press.

Van Volkenburgh, E. and Boyer, J.S. (1985) Inhibitory effects of water deficit on maize leaf elongation. *Plant Physiology* 77, 190-194.

Wilkinson, S. and Davies, W.J. (1997) Xylem sap pH increase: a drought signal received at the apoplastic face of the guard cell that involves the suppression of saturable abscisic acid uptake by the epidermal symplast. *Plant Physiology* 113, 559-573.

Wu, Y., Spollen, W.G., Sharp, R.E., Hetherington, P.R. and Fry, S.C. (1994) Root growth maintenance at low water potentials. Increased activity of xyloglucan endotransglycosylase and its possible regulation by abscisic acid. *Plant Physiology* 106, 607-615.

Zhang, J. and Davies, W.J. (1990) Does ABA in the xylem control the rate of leaf growth in soil-dried maize and sunflower plants? *Journal of Experimental Botany* 41, 1125-1132.

Zhang, J. and Davies, W.J. (1991) Antitranspirant activity in xylem sap of maize plants. *Journal of Experimental Botany* 42, 317-321.

14 Engineering Phytochrome Genes to Improve Crop Performance

H. SMITH

Department of Biology, University of Leicester, Leicester LE1 7RH, UK

Crowd Trouble in Plants

When plants grow in dense communities, as in crop stands, the availability of light energy for photosynthesis is potentially limiting. Plants have evolved two contrasting strategies that enhance survival under conditions of competition for light. Shade tolerance, a combination of mechanisms that improves the biochemical efficiency of light capture whilst reducing the expenditure of energy through respiratory activity, is employed by plants that are genetically adapted to shaded conditions. Shade tolerance is common among the vascular cryptogams, whose energy requirements for growth and development are relatively low, and whose demands on photosynthetic energy acquisition are correspondingly conservative. A high efficiency of light capture and utilization allows such plants to survive often in deeply shaded environments. Some higher plants assume this strategy, particularly those that we often grow as house plants, as these are usually plants adapted to the dense shade of tropical and subtropical forests. For most of the angiosperms, however, coping with shade is a matter of avoidance, rather than tolerance. Shade avoidance is a syndrome of developmental responses that collectively allow plants to maintain photosynthetic structures in light conditions that are not limiting for growth. Shade-avoiding plants have high resource capture requirements, and generally are profligate with the utilization of those resources. In other words, shade avoiders are aggressive, competitive plants whose existence depends on the ability to dominate in the battle for the resource of light. Almost all the crop plants utilized in modern, intensive farming are strong shade avoiders.

That plants grow long and slender when grown in dense stands has probably been known for centuries. The first reference I have found is by that great English botanist, Stephen Hales, who wrote in his *Statical Essays*, in 1727:

Beans and many other plants, which stand where they are much shaded, being thereby kept continually moist, do grow to unusual heights, and are drawn up as they call it, by the overshadowing trees, their parts being kept long, soft and ductile.

This is a clear description of shade avoidance, with emphasis on increased extension growth, and in the 'soft and ductile' nature of the resulting plant. Shade avoidance is a syndrome of physiological responses (see Table 14.1) initiated by signals in the light environment that are unique to situations of crowded vegetation. For most of the time since Stephen Hales drew attention to this phenomenon, it generally has been assumed that such plants were 'growing towards the light'. In fact, we now know that plants grow away from the shade, rather than towards the light.

Shade avoidance involves the perception of specific light signals reflected from neighbours. The photoreceptors responsible are the phytochromes, and they have been supremely designed through evolution to detect the reflection of radiation from nearby vegetation. The phytochromes are unique amongst bio-

Table 14.1. The shade avoidance syndrome.

Physiological process	Response to shade (i.e. reduced R:FR ratio)
Germination	Retarded
Extension growth	Accelerated
Internode extension	Rapidly increased (lag c. 5 min)
Petiole extension	Rapidly increased
Leaf extension	Increased in cereals
Leaf development	Retarded
Leaf area growth	Marginally reduced
Leaf thickness	Reduced
Chloroplast development	Retarded
Chlorophyll synthesis	Reduced
Chlorophyll a:b ratio	Balance changed
Apical dominance	Strengthened
Branching	Inhibited
Tillering (in cereals and grasses)	Inhibited
Flowering	Accelerated
Rate of flowering	Markedly increased
Seed set	Severe reduction
Fruit development	Truncated
Senescence	Accelerated
Leaf senescence	Advanced
Leaf abscission	Advanced
Assimilate distribution	Marked change
Storage organ deposition	Severe reduction

logical photoreceptors as they respond to the relative amounts of two radiation signals, rather than to the absolute amount of a single signal. The phytochromes are chromoproteins with linear tetrapyrrole chromophores existing in two forms that either absorb red (R) or far-red (FR) light, each of which causes conversion to the other form, according to the following scheme:

$$\text{Pr} \underset{\text{Far-red}}{\overset{\text{Red}}{\rightleftarrows}} \text{Pfr} \longrightarrow \text{Biological action}$$

These photochemical properties endow the phytochromes with the capacity to measure the relative amounts of red and far-red radiation incident upon the plant. Perception of the R:FR comparative signal provides the plant with unambiguous information on the proximity of neighbours. In this chapter, I wish to show how recent knowledge of the phytochromes has led to a new genetic engineering strategy to improve crop plant performance.

Phytochromes and Light Signal Perception

The phytochromes are a family of chromoprotein photoreceptors (Clack et al., 1994) that, as far as is known, all have similar chromophores. The signal perception capacities of the phytochromes are determined by their absorption characteristics, and by the kinetic relationships between the two stable forms of the molecule, i.e. Pr and Pfr.

Figure 14.1 shows the absorption spectra of phytochrome purified from etiolated oat seedlings taken after far-red radiation (i.e. Pr) and after red irradiation (i.e. predominantly Pfr). Pr absorbs strongly in the red, with an absorption maximum (λ_{max}) at approximately 665 nm, and Pfr absorbs maximally in the far-red with a broader λ_{max} at approximately 730 nm. A very important consequence of the broad absorption spectra of Pr and Pfr is that, under normal radiation conditions in the natural environment, the two forms of phytochrome are being continually interconverted. The proportion of the total population of molecules present at any one instant in the active Pfr form (denoted by Pfr/P) is determined by the absorption coefficients of the two forms and by the spectral photon distribution of the radiation. In this way, the phytochromes act as continual sensors of the spectral distribution of radiation in the 600–800 nm waveband. In principle, the photoequilibrium between Pr and Pfr can be calculated from the absorption cross-sections of Pr and Pfr, and the spectral photon flux distribution of the actinic radiation. In practice, it is simpler to relate Pfr/P to the relative amounts of red and far-red (i.e. the R:FR ratio) in the actinic radiation (Smith and Holmes, 1977). Conventionally, the R:FR ratio is measured as the ratio of photon flux in 10-nm wide bands centred on 660 and 730 nm, and these data can be obtained readily from a spectroradiometric scan. By demonstrating that a particular biological response is a function of the R:FR ratio, and therefore of Pfr/P, it is possible to show that the response is mediated by one or more of the phytochromes. Such R:FR ratio responses provide the

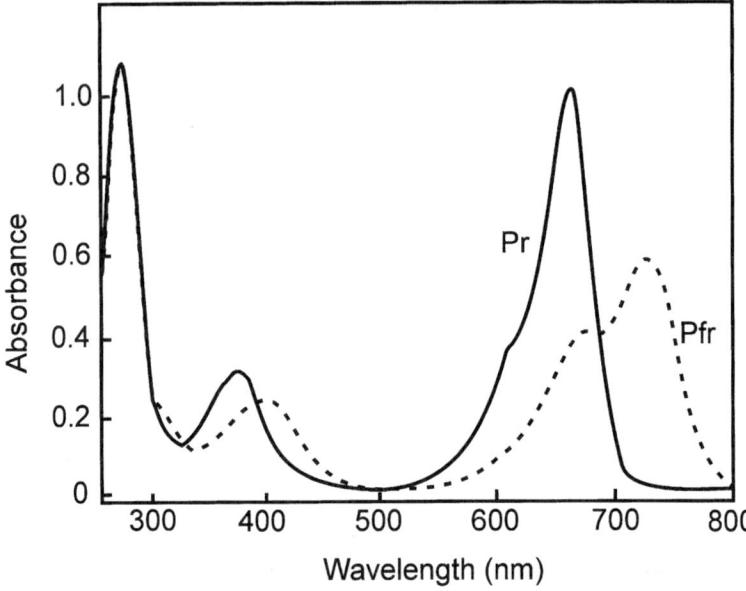

Fig. 14.1. The absorption spectra of the phytochromes. In the Pr form, the absorption maximum of the phytochromes is close to 665 nm, whereas in the Pfr form it is shifted to higher wavelengths by about 60 nm to 730 nm. Absorption of a photon by a Pr molecule converts Pr to Pfr; absorption of a photon by a Pfr molecule converts Pfr to Pr. Both Pr and Pfr absorb light at wavelengths below about 730 nm, so in broad-band radiation such as daylight, both photoconversions occur simultaneously, leading to a photoequilibrium. At equilibrium, the steady-state proportions of Pr and Pfr are determined principally by the relative amounts of red and far-red radiation, i.e. the R:FR ratio.

experimental basis for concluding that shade avoidance responses are mediated by the phytochromes (Smith, 1982).

Phytochromes in the Natural Environment

The relationship between the R:FR ratio of natural or artificial broadband radiation and Pfr/P established by that radiation is both experimentally, and theoretically, a rectangular hyperbola (Fig. 14.2). The R:FR ratio of unfiltered solar radiation (with a solar angle of $> 10°$) approximates to 1.12, and is largely unaffected by random fluctuations in the total amount of radiation caused by weather conditions (Holmes and Smith, 1977). This value of the R:FR ratio establishes a photoequilibrium of about 0.55, which is on the shoulder of the hyperbola in Fig. 14.2, and represents a norm against which the plant may compare light signals of potentially important environmental conditions. The R:FR ratio potentially is a signal of a number of important environmental conditions, including depth underwater, the day–night or night–day transition,

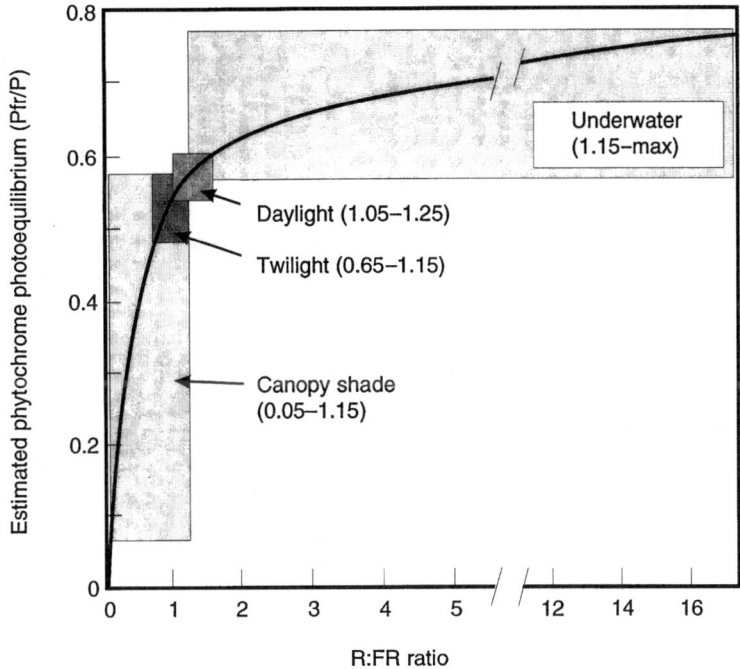

Fig. 14.2. The relationship between R:FR ratio and the phytochrome photoequilibrium. The acute sensitivity of the phytochromes as shade and/or proximity sensors is demonstrated by the relationship between R:FR and Pfr/P. In daylight, R:FR is very constant within the range of about 1.05–1.25. Phytochrome exposed to daylight establishes approximately 55% of the total in the Pfr form, and correspondingly 45% in the Pr form, although photoconversions between the two forms proceed at a rapid rate. The relationship between R:FR and Pfr/P is a rectangular hyperbola, and the region of the steepest part of the curve coincides with the R:FR ratios that occur within plant canopies. Other environmental conditions affect R:FR (e.g. at dawn and dusk, and underwater), but the shape of the relationship shows that the phytochromes are extremely sensitive detectors of vegetation reflection signals.

and the proximity of neighbouring vegetation. For good reasons (not developed here, but see Smith, 1982, for review), however, it is likely that only the latter signal is of real significance.

Vegetation is characterized by high concentrations of the photosynthetic pigments, which absorb radiation very strongly in the red and blue wavebands. Whenever light interacts with vegetation, therefore, there is a large change in spectral distribution, with transmitted or scattered radiation being deficient in red and blue wavelengths, and relatively (sometimes absolutely) enhanced in the far-red. These changes in R:FR occur on the steep slope of the rectangular hyperbola and, therefore, potentially could act as highly sensitive signals of shade, or impending shade. A small reduction in R:FR from that of the daylight norm causes a relatively large change in Pfr/P, the exact requirement of a sensitive detection system.

Phytochromes as Proximity Sensors

Within natural vegetation stands, both the amount of radiation and the spectral distribution of that radiation vary as a function of the proximity and mass of vegetation with which the radiation interacts (Gilbert *et al.*, 1995). In nature, therefore, it is impossible to determine whether the developmental responses that comprise shade avoidance are caused by changes in irradiance or in spectrum. In controlled conditions, on the other hand, it is possible to reduce the R:FR ratio not by diminishing the radiation in the red wavelengths, but by elevating the radiation in the far-red spectral region. If shade avoidance is caused by phytochrome-mediated perception of reduced R:FR, then shade avoidance responses should be evoked under such artificial conditions. In other words, by increasing the total amount of radiation through far-red supplementation, shade avoidance responses should be observed.

Figure 14.3 shows elongation growth data for *Chenopodium album* plants growing in far-red-supplemented white light. These plants are aggressive weeds, and the most striking response is the increase in stem elongation seen at low R:FR ratios. Surprisingly, the relationship between Pfr/P and elongation rate was linear (Morgan and Smith, 1976). The slope of the relationship between growth rate and Pfr/P differs between species, of course, but the variability is systematic. Since these early observations, many species have been tested and all show qualitatively similar responses, indicating that R:FR ratio detection is universal amongst plant species. Species that are genetically adapted to shaded habitats, and whose normal strategy is shade tolerance, show shallow slopes, whereas aggressive, competitive plants show steep slopes (Morgan and Smith, 1979). Most importantly, many crop species have now been tested, and all have displayed strong extension growth responses to supplementary far-red radiation. Species tested include: wheat, barley, maize, sunflower, flax, sugar beet, radish, tomato and tobacco (mostly unpublished data of the Leicester group).

Nature of Shade Avoidance

Shade avoidance responses are, by definition, those responses that occur when plants are grown in close proximity. They include increased stem and petiole elongation, reduced leaf development and accelerated senescence, elevated leaf angle, enhanced apical dominance and reduced branching, accelerated flowering and fruit production, reduced seed set and, overall, a redistribution of assimilates from leaf and storage organ development towards extension growth (Table 14.1). All of these phenomena can be observed under natural shade avoidance conditions, and can be simulated by growing plants in cabinets in which the R:FR ratio is reduced by supplementation of white light with far-red.

Shade avoidance represents a systematic modification of the developmental pattern, in response to a simple environmental signal. In this respect, shade avoidance exemplifies the plasticity of plant development (Schmitt and Wulf, 1993). Although, obviously, shade avoidance is genetically determined, the

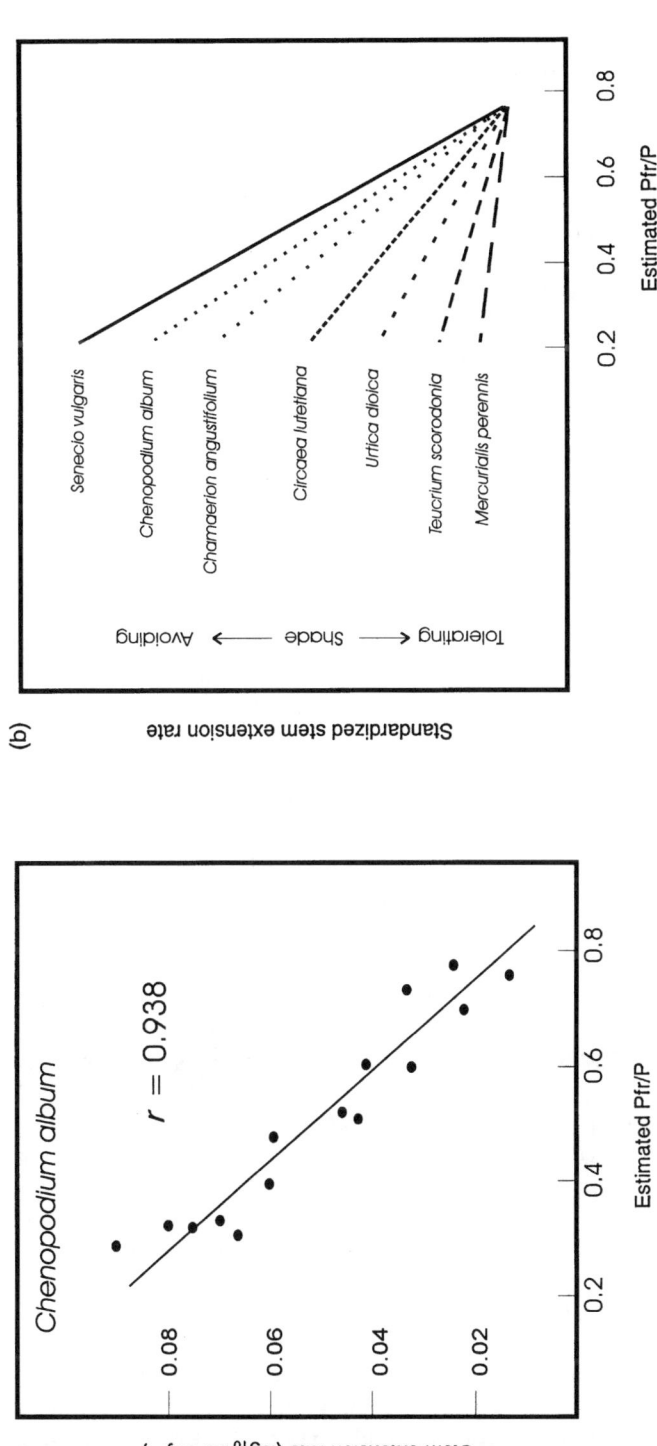

Fig 14.3. (a) Decreasing the phytochrome photoequilibrium causes linear increases in extension growth rate. *Chenopodium album* were grown in growth cabinets in which the total amount of photosynthetically active radiation (PAR = 400–700 nm) was held constant and uniform, but in which different amounts of far-red radiation were added to vary the R:FR. The data are plotted here as stem extension rate versus estimated phytochrome photoequilibrium (Pfr/P). (b) The strength of shade avoidance varies systematically with ecological habitat. All plants that have been tested using the R:FR ratio cabinets show relationships similar to that for *Chenopodium*. The strength of the response to a low R:FR ratio depends on the genetic adaptation of the species (or ecotype) to shaded or open habitats. Species adapted to open habitats, such as the aggressive weeds at the top of this figure, display marked responses, whereas those adapted to shaded habitats still respond, but do so relatively weakly. In this figure, the stem elongation rates of the various species, which of course are in absolute terms very different, have been normalized to the rate at the highest R:FR ratio used, in order to facilitate comparisons. (Data from Morgan and Smith, 1976, 1979.)

capacity of the plant to modify its developmental pattern in response to environmental signals is nowhere more spectacular. It is plasticity that allows individual plants to react to environmental stimuli in such a way that they can maximize their chances of survival. In the crop situation, on the other hand, competition between individuals within the crop is counterproductive, from the farmer's viewpoint. Each individual plant in a field will respond to the signals reflected or scattered from its neighbours, resulting in often quite marked reallocation of resources from leaf and storage organ production to stem and petiole elongation. These normal shade avoidance responses have the effect of reducing the potential harvest index, and also have undesirable secondary effects of increased susceptibility to lodging.

Shade avoidance is a crucial strategy for competition in the natural environment, but it is bad news for crop performance!

Roles of Individual Phytochromes in Shade Avoidance

The phytochromes are a family of photoreceptors, consisting of five members (phytochromes A, B, C, D and E) in the angiosperms (Clack et al., 1994; Matthews and Sharrock, 1997). An important goal is to determine which of the phytochromes is responsible for the perception of R:FR ratio signals and the induction of shade avoidance responses. Armed with such knowledge, it should be possible to mount programmes to manipulate shade avoidance to the advantage of agricultural production. The most compelling evidence comes from studies of photomorphogenic mutants.

There are many mutants that are deficient in specific aspects of photomorphogenesis. Photomorphogenic mutants are defined as mutants lacking specific light-mediated responses, and can be classified into three main groups: photoreceptor mutants, chromophore synthesis mutants and transduction chain mutants. The most useful for elucidating the functions of individual members of the phytochrome family are the photoreceptor mutants, in which the lesion is within the gene encoding the photoreceptor apoprotein. Mutants in this category are available in several plants, with *Arabidopsis* providing the most complete range. *Arabidopsis* mutants are now available that have lesions in the genes encoding phytochromes A, B, D and E. Some alleles of these mutants are completely devoid of the specific members of the family in which the mutation is located, and therefore provide excellent material for assessing the roles of individual phytochromes in specific responses. The allocation of function to phytochrome family members is an ongoing project at the time of writing, but a number of conclusions can be reached from studies of the *Arabidopsis* mutants (see Whitelam and Harberd, 1994; Smith 1995; Whitelam and Devlin, 1997).

Taken collectively, the evidence from mutants is that the shade avoidance syndrome is probably mediated by multiple phytochromes (Smith and Whitelam, 1997). Phytochrome B is a very important regulator of cell elongation in etiolated and in light-grown plants, and it is clear that it has a major function in shade avoidance. However, the mutants have shown that phytochrome B does not act alone, nor is it sufficient for full shade avoidance. Current thinking implicates both phytochromes D and E in shade avoidance, with perhaps phytochrome E playing a more important supporting role than phytochrome D. It may turn out that individual components of the shade avoidance syndrome are mediated by individual phytochromes, but at present it is more prudent to think in terms of overlapping functions.

Another very important outcome of intensive studies of phytochrome mutants is the apparent antagonism of phytochromes A and B in the regulation of stem elongation (Smith *et al.*, 1997). Although phytochrome A is the predominant member of the family in plants grown in the dark, as seedlings de-etiolate, the concentration of phytochrome A diminishes rapidly, with a half-life often of less than 30 min. For this reason, phytochrome A appears to have only a minor role in light-grown plants (except in the photoperiodic control of flowering, see Bagnall *et al.*, 1995). Mutants lacking phytochrome A have a wild-type phenotype when grown in the light (Whitelam *et al.*, 1993), and therefore there can only be a minor role, if any, for phytochrome A in shade avoidance. The realization that phytochrome A action can antagonize that of phytochrome B, however, gave a clue to a potential approach to the suppression of shade avoidance through genetic engineering.

Genetic Engineering of Shade Avoidance

The isolation of genes encoding the phytochromes (i.e. *PHY* genes) has made possible an approach towards the modification of plant responses to light signals based upon transgenic expression. Both sense and antisense expression of *PHY* sequences have been reported, for several source genes and host plants (Robson and Smith, 1997). The first *PHY* cDNAs to be isolated were those encoding phytochrome A in oats and rice, and transgenic expression of the oat *PHYA* cDNA, driven by the 'constitutive' cauliflower mosaic virus (CaMV) 35S promoter, in both tobacco and tomato yielded extremely dwarf, dark-green plants (Boylan and Quail, 1989; Keller *et al.*, 1989). Later, *PHYB* cDNA expression also yielded dwarf plants, both in tobacco and *Arabidopsis*, although the extent of the phenotype varied depending on the source of the cDNA (i.e. *Arabidopsis* or rice) (Wagner *et al.*, 1991; McCormac *et al.*, 1993).

Overexpression of *PHYA* sequences has, to date, been the principal source of potential improvements to crop plant performance. Observations of transgenic tobacco expressing an oat *PHYA* cDNA to relatively high levels revealed an entirely unexpected result; these plants exhibited *negative* shade avoidance (McCormac *et al.*, 1991, 1992). In other words, whereas the wild-type plants elongated much more rapidly in white light with added far-red (i.e. W + FR) than they did in white light alone, the transgenic plants were dwarfed under W + FR conditions. The explanation for this initially bizarre result requires an elaboration of what is known about the properties of phytochrome A.

Phytochrome A is very different from the other members of the phytochrome family. It is synthesized in relatively high quantities in dark-grown seedlings, and reaches comparatively high levels in etiolated seedlings. Almost everything known about the chemical properties of the phytochromes therefore comes from analyses of phytochrome A. Once a plant is transferred to the light, phytochrome A is degraded rapidly and, furthermore, its synthesis is strongly down-regulated. The half-life of the loss of phytochrome A after transfer to the light is very short, in some cases as short as 15–20 min. Thus, after 12–24 h, there is very little phytochrome A left in the cells, and indeed the concentration of phytochrome A at steady state in a light-grown plant is much less than that of phytochrome B. Phytochrome B and, by extension, the other phytochromes are not subject to such degradation in the light, and are therefore known as light-stable phytochromes. This means that the physiological role of phytochrome A is very important in seedlings that are undergoing de-etiolation, but its function is relatively unimportant in the mature, light-grown plant.

Phytochrome A is commonly described as being 'light-labile' but, like much other scientific jargon, this description is not strictly correct. What happens is that the Pfr form of phytochrome A (i.e. Pfr_A) is very unstable; thus after Pfr_A is formed from Pr_A by the action of light, it is degraded rapidly, whether or not the light treatment is maintained. If, however, the light treatment is extended, the continual photoconversion of the residual Pr_A to Pfr_A results in the loss of total phytochrome A down to very low levels, at which the very slow synthesis is matched by degradation at steady state. The overall dynamics are shown by the following scheme:

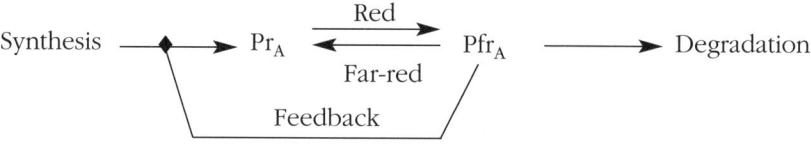

As stated above, the dynamic changes in phytochrome A concentration upon exposure to light mean that this phytochrome has only a temporary action on development of the normal, wild-type plant. Phytochrome A mediates the so-called 'high irradiance response', or HIR. The HIR requires long-term irradiation and has an action maximum at a wavelength in the far-red intermediate between the absorption maxima of Pr and Pfr (Hartmann, 1966). The FR-HIR has been the subject of considerable controversy for more than 30 years, and even now the mechanism through which far-red light can cause strong inhibition of hypocotyl extension is not fully understood. It is certain, however, that phytochrome A mediates the FR-HIR, and that the mechanism is at least partly dependent upon the instability of Pfr_A. Wavelengths that establish high proportions of Pfr_A lead to the rapid loss of total phytochrome A and are biologically ineffective, but wavelengths intermediate between 660 and 730 nm, that establish low proportions of Pfr_A are very effective. The rationale is that low concentrations of Pfr_A, if present for a long time, maximize the action. This explanation is intellectually satisfying, as it is consistent with our knowledge of the properties of phytochrome A, but other aspects of the FR-HIR (fortunately outside the scope of this chapter!) are more difficult to account for.

The role of phytochrome A in inhibiting extension growth under con-

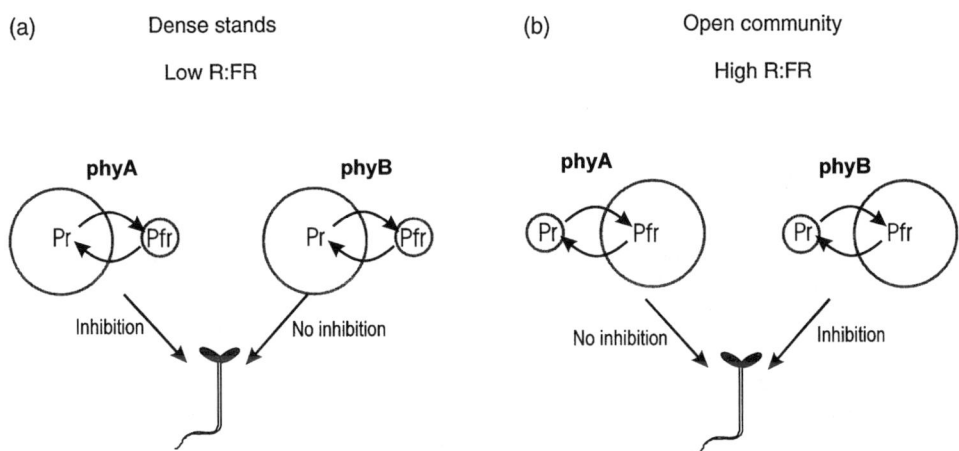

Fig. 14.4. Scheme for the apparent antagonism between phytochrome A and phytochrome B. In high R:FR ratio light, such as open daylight (b) the photoequilibrium favours Pfr, and under these circumstances phytochrome B (phyB) mediates an inhibition of extension growth, whereas phyA has little effect. Under dense canopy conditions with a low R:FR ratio (a) phyA, operating through the FR-HIR (see text) inhibits extension growth, whereas phyB is ineffective. In this diagram, the diameter of the circles for Pr and Pfr is intended to indicate the relative pool sizes. This diagram illustrates both the situation in wild-type seedlings during de-etiolation, when both phytochrome A and phytochrome B are present and active, and in mature transgenic plants expressing a *PHYA* gene. Thus, the antagonism between phytochrome A and phytochrome B is used in transgenic plants to suppress the shade avoidance responses mediated by phytochromes B, D and E.

tinuous far-red light in fact accounts for the apparent antagonism of phytochromes A and B alluded to above (Smith *et al.*, 1997). During the period in which both phytochromes A and B are active in the de-etiolating seedling, this antagonism manifests itself as a conflict between opposing effects of supplementary far-red light. Acting through phytochrome A, far-red inhibits extension growth via the FR-HIR whereas, through phytochrome B, far-red light releases the Pfr$_B$-mediated extension growth via the shade avoidance reactions. Figure 14.4 presents a scheme to illustrate the way in which phytochrome A and phytochrome B antagonism operates in the natural environment. Such antagonism may also be deduced from mutants that lack phytochrome A; these mutants exhibit a stronger stimulation of extension growth in white light supplemented with far-red, than do comparative wild-type seedlings. These data show that phytochrome A still manifests some regulation of extension growth in light-grown plants, but in the wild-types the level of control is minimal. The observations, however, suggest that if phytochrome A concentrations could be engineered to remain high after de-etiolation, then the persistence of a FR-HIR might counteract the elongation responses mediated by phytochrome B (and other phytochromes), thereby eliminating or suppressing shade avoidance responses.

In seedlings overexpressing a transgenic *PHYA* cDNA, the dynamics of phytochrome A concentration are expected to be very different. On the one hand, expression driven by the CaMV 35S promoter will be independent of the feedback regulation that down-regulates the activity of the endogenous phytochrome A promoter. Furthermore, the mechanism of phytochrome A degradation, known to be via the ubiquitin pathway, may be sufficiently specific that the heterologous phytochrome A may be spared degradation. Both of these predictions have been fulfilled. In transgenic tobacco expressing an oat *PHYA* cDNA, for example, the half-life of the heterologous phytochrome A is close to 4 h, whereas that of the native phytochrome A is less than 1 h (McCormac et al., 1992). Thus, the heterologous phytochrome A persists through de-etiolation and, indeed in the transgenic tobacco mentioned above, the existence of a strong phenotype in adult plants demonstrates that the phytochrome A remains at sufficient concentration to be active throughout the life of the plant. Overexpressed phytochrome A, therefore, is able to maintain a FR-HIR after the plants have de-etiolated, resulting in the suppression of normal shade avoidance responses to supplementary far-red, and indeed to the negative shade avoidance response described above. This being so, the opportunity therefore exists to exploit the transgenic expression of phytochrome A to suppress shade avoidance in the field.

Transgenic Suppression of Shade Avoidance in the Field

Suppression of shade avoidance can be expected to provide a number of advantages for crop plants grown at high densities (Smith, 1992). Virtually all known crops are strong shade avoiders, and plant breeding through the centuries has apparently not bred out elongation responses to crowding. Even the dwarf cereals that have led to marked improvements in agricultural productivity via the so-called 'Green Revolution' still show very strong shade avoidance responses. As most crops are grown at high density, the resulting shade avoidance causes elongated stems and petioles, reduced leaf development, accelerated flowering and reduced seed set, and, overall, the allocation of resources to parts of the plant that are not harvested. Shade-avoiding crops grow long and have weak stems, making them susceptible to lodging. In the horticultural industry, massive amounts of chemical growth regulators are applied to curtail stem elongation. For example, in the cultivation of poinsettias for seasonal retail at Christmas, it is not uncommon for each plant to receive more than 30 applications of Cycocel, or some other antagonist or inhibitor of gibberellin action. Suppression of shade avoidance by a means that is not environmentally damaging would be a considerable breakthrough.

Field experiments using tobacco as a model crop plant have confirmed that transgenic expression of a *PHYA* cDNA suppresses shade avoidance in the field (Robson *et al.*, 1996). For these experiments to be successful, it was necessary to select transgenic lines that expressed the heterologous cDNA to intermediate levels, since high-level expression caused unacceptable dwarfing. It was found that expression causing the concentration of phytochrome A in etiolated plants

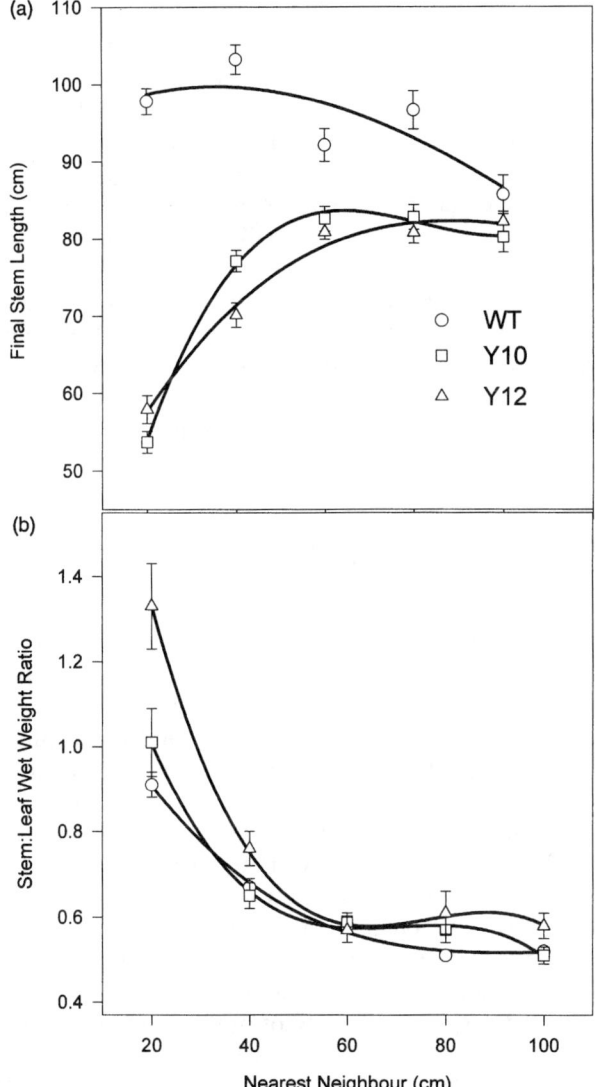

Fig. 14.5. Proximity-conditional dwarfing in the field. This figure shows the final harvest data for a field experiment in which wild-type and transgenic tobacco were grown at a range of planting densities. The transgenic strains are called Y10 and Y12. At a sparse planting density of 1 m, all three strains reached similar final plant heights (a). As density was increased, the wild-type plants showed a certain degree of shade avoidance, but the transgenic *PHYA* expressers exhibited negative shade avoidance. Thus, these plants were dwarfed only when they were grown in close competition with neighbours – in other words, the ectopic phytochrome A causes proximity-conditional dwarfing. In part (b) the ratio of stem to leaf biomass is shown. As density increased, the wild-type plants committed an increasing larger proportion of resources to stem growth, at the expense of leaf production. The same trend was observed in the transgenic plants, but it was markedly reduced, leading to an increased yield of leaf material. (Data from Robson *et al.*, 1996.)

to be about two- to threefold that of the native phytochrome A did not produce dwarfing in ordinary daylight. However, when the plants were grown at high densities, then the negative shade avoidance phenomenon described above caused the plants to be significantly dwarfed compared with the wild-types. Figure 14.5a shows the harvest data for the mean height of the wild-type and transgenic plants as a function of the proximity of neighbours. Strong negative shade avoidance is seen in the two transgenic lines. Furthermore, the allocation of resources to stem growth was markedly reduced in the transgenic lines at high density, in comparison with the wild-types (Figure 14.5b). If harvest index in tobacco is defined as the proportion of total biomass that is composed of leaf tissue, then transgenic expression of the oat *PHYA* cDNA increased harvest index by up to 20% at high densities. We do not know yet whether such impressive improvements in performance will be obtained with more conventional crops, but at the time of writing a major effort is being undertaken worldwide to transfer this approach to other crops.

Crystal-balling About *PHY* Transgenes

The phytochromes represent an impressive armoury of photoreceptors that collectively regulate almost all aspects of development in the natural environment. The more one learns about the discrete and the overlapping functions of the different phytochrome family members, the more one becomes fascinated by their plurality and flexibility. The action of individual phytochromes is a function of the concentration of that phytochrome – this much is obvious from the semi-dominant nature of the photoreceptor mutations. In consequence, manipulation of the concentration of individual phytochromes, through genetic engineering, will be expected to lead to severe effects on phenotype. In principle, therefore, many agronomically important traits may be directly modifiable through *PHY* gene engineering. Furthermore, it now seems likely that each of the phytochromes operates through distinct transduction pathways, leading to the speculation that manipulation of transduction chain components may prove to be a very valuable way of modifying plant performance in the field.

Acknowledgements

The research cited in this chapter was funded variously by the Natural Environment Research Council, The Biotechnology and Biological Sciences Research Council and the Commission of the European Community

References

Bagnall, D., King, R.W., Whitelam, G.C., Boylan, M.T., Wagner, D. and Quail, P.H. (1995) Flowering responses to altered expression of phytochrome in mutants and transgenic lines of *Arabidopsis*. *Plant Physiology* 108, 1495-1503.

Boylan, M.T. and Quail, P.H. (1989) Oat phytochrome is biologically active in transgenic tomatoes. *The Plant Cell* 1, 765-773

Clack, T., Mathews, S. and Sharrock, R.A. (1994) The phytochrome apoprotein family in *Arabidopsis* is encoded by five genes: the sequences and expression of *PHYD* and *PHYE*. *Plant Molecular Biology* 25, 413-427.

Gilbert, I.R., Seavers, G.P., Jarvis, P.G. and Smith, H. (1995) Photomorphogenesis and canopy dynamics. Phytochrome-mediated proximity perception accounts for the growth dynamics of canopies of *Populus trichocarpa* × *deltoides* 'Beaupré'. *Plant, Cell and Environment* 18, 475-497.

Hartmann, K.M. (1966) A general hypothesis to interpret 'high energy phenomena' of photomorphogenesis on the basis of phytochrome. *Photochemistry and Photobiology* 5, 349-366.

Holmes, M.G. and Smith, H. (1977) The function of phytochrome in the natural environment. I. Characterisation of daylight for studies in photomorphogenesis and photoperiodism. *Photochemistry and Photobiology* 25, 533-538.

Keller, J.M., Shanklin, J., Vierstra, R.D. and Hershey, H.P. (1989) Expression of a functional monocotyledonous phytochrome in transgenic tobacco. *EMBO Journal* 8, 1005-1012

Matthews, S. and Sharrock, R.A. (1997) Phytochrome gene diversity. *Plant, Cell and Environment* 20, 666-671.

McCormac, A.C., Cherry, J.R., Hershey, H.P., Vierstra, R.D. and Smith, H. (1991) Photoresponses of transgenic tobacco plants expressing an oat phytochrome gene. *Planta* 185, 162-170

McCormac, A.C., Whitelam, G.C. and Smith, H. (1992) Light-grown plants of transgenic tobacco expressing an introduced oat phytochrome A gene under the control of a constitutive viral promoter exhibit persistent growth inhibition by far-red light *Planta* 188, 173-181.

McCormac, A.C., Wagner, D., Boylan, M.T., Quail, P.H., Smith, H. and Whitelam, G.C. (1993) Photoresponses of transgenic *Arabidopsis* seedlings expressing introduced phytochrome B-encoding cDNAs: evidence that phytochrome A and phytochrome B have distinct photoregulatory functions. *The Plant Journal* 4, 19-27.

Morgan, D.C. and Smith, H. (1976) Linear relationship between phytochrome photoequilibrium and growth in plants under simulated natural radiation. *Nature* 262, 210-212.

Morgan, D.C. and Smith, H. (1979) A systematic relationship between phytochrome-controlled development and species habitat for plants grown in simulated natural radiation. *Planta* 145, 253-259.

Robson, P.R.H. and Smith, H. (1997) Fundamental and biotechnological applications of phytochrome transgenes. *Plant, Cell and Environment* 20, 831-839.

Robson, P.R.H., McCormac, A.C., Irvine, A.S. and Smith, H. (1996) Genetic engineering of harvest index in tobacco through overexpression of a phytochrome gene. *Nature Biotechnology* 14, 995-998

Schmitt, J. and Wulf, R.D. (1993) Light spectral quality, phytochrome, and plant competition. *Trends in Ecology and Evolution* 8, 47-51.

Smith, H. (1982) Light quality, photoperception and plant strategy. *Annual Review of Plant Physiology* 33, 481-518.

Smith, H. (1992) Ecology of photomorphogenesis: clues to a transgenic programme of crop plant improvement. *Photochemistry and Photobiology* 56, 815-822.

Smith, H. (1995) Physiological and ecological function within the phytochrome family. *Annual Review of Plant Physiology and Plant Molecular Biology* 46, 289-315

Smith, H. and Holmes, M.G. (1977) The function of phytochrome in the natural environment. III. Measurement and calculation of phytochrome photoequilibrium. *Photochemistry and Photobiology* 25, 547-550.

Smith, H. and Whitelam, G.C. (1997) The shade avoidance syndrome: multiple responses mediated by multiple photoreceptors. *Plant, Cell and Environment* 20, 840-844.

Smith, H., Xu, Y. and Quail, P.H. (1997) Antagonistic but complementary actions of phytochromes A and B allow optimum seedling de-etiolation. *Plant Physiology* 114, 637-641.

Wagner, D., Tepperman, J.M. and Quail, P.H. (1991) Overexpression of phytochrome-B induces a short hypocotyl phenotype in transgenic *Arabidopsis*. *The Plant Cell* 3, 1275-1288.

Whitelam, G.C. and Devlin, P.F. (1997) Roles of different phytochromes in *Arabidopsis* photomorphogenesis. *Plant, Cell and Environment* 20, 752-758.

Whitelam, G.C. and Harberd, N.P. (1994) Action and function of phytochrome family members revealed through the study of mutant and transgenic plants. *Plant, Cell and Environment* 17, 615-625.

Whitelam, G.C., Johnson, E., Peng, J., Carol, P., Anderson, M.L., Cowl, J.S. and Harberd, N.P. (1993) Phytochrome A null mutants of *Arabidopsis* display a wild-type phenotype in white light. *The Plant Cell* 5, 757-768.

15 Regulation of Stem Extension by Temperature

F.A. LANGTON

Horticulture Research International, Wellesbourne, Warwick CV35 9EF, UK

Introduction

The regulation of stem extension is particularly important in the cultivation of greenhouse ornamental plants, since these have to meet strict quality specifications for height and overall size. For much of the year, the need is to make plants more compact, and this is generally achieved by the use of plant growth regulator chemicals. It is not unusual, for example, for poinsettia (*Euphorbia pulcherrima*) crops grown in the UK to receive 15 or more sprays of chlormequat (Cycocel). However, the application of chemical sprays is costly, and the chemicals themselves increasingly are being perceived as environmentally unfriendly. Thus, there is growing interest in alternatives to chemical control, and the manipulation of temperature to regulate height is already being widely practised. The problem is that the physiological principles which underlie the effects of temperature on extension growth are not fully understood, and this leads to uncertainties over how best to optimize the use of temperature.

Day–Night Temperature Responses

It has long been known that plants of most species grown with a high day temperature (DT) and a low night temperature (NT) are tall with long internodes, whilst plants grown in the reciprocal regime, low DT-high NT, are compact with short internodes. This is exemplified in Fig. 15.1a which shows tomato (*Lycopersicon esculentum*) plants 14 days after transfer to either a high DT (25°C)-low NT (15°C) regime (left) or to a low DT (15°C)-high NT (25°C) regime (right) (D = N = 12 h). Response to change in temperature is extremely rapid, and Fig. 15.1b shows the same two plants 14 days later, with the high DT-low NT plant having been transferred to the low DT-high NT regime (left), and the low DT-high NT plant having been transferred to the high DT-low NT

© CAB INTERNATIONAL 1998. *Genetic and Environmental Manipulation of Horticultural Crops* (eds K.E. Cockshull, D. Gray, G.B. Seymour and B. Thomas)

Fig. 15.1. Tomato plants, cv. Moneymaker, growing in reciprocal day temperature (DT)–night temperature (NT) regimes (D = N = 12 h): (a) 14 days after transfer of plants from continuous 20 to 25°C DT–15°C NT (left) or 15°C DT–25°C NT (right); (b) the same two plants, 14 days later, after transfer to the reciprocal regime: 25°C DT–15°C NT transferred to 15°C DT–25°C NT (left); 15°C DT–25°C NT transferred to 25°C DT–15°C NT (right).

regime. Extension growth in this second 14-day period can be seen to have been such that the plants are again of a similar height. The extreme responsiveness and the ease with which tomato can be grown make this species a frequently used 'model' in stem extension studies.

Responses of the sort illustrated in Fig. 15.1a have been reported previously for tomato by Heuvelink (1989). He showed that young plants grown for 40 days with DT lower than NT (16°C DT–26°C NT) were 56% shorter than plants grown in the reciprocal regime (26°C DT–16°C NT), but that leaf production was reduced by only 15%. This indicates that effects on leaf emergence rate play little part in explaining the differences in height given by such reciprocal temperature regimes. Furthermore, differences in height due to effects on leaf production take longer to manifest themselves than the 2 weeks required to give the differences shown in Fig. 15.1. The rapidity of response also rules out effects mediated directly via photosynthesis as a primary cause although, in the longer term, large reductions in plant dry weight can be expected for plants grown in the low DT–high NT regime (Heuvelink, 1989). Factors generally associated with low DT–high NT regimes which contribute to reduced dry weight include smaller leaves and reduced chlorophyll per unit leaf area (Erwin and Heins, 1995), but these are generally perceived to be of only minor consequence in the production of ornamental species, especially since plants rapidly 'green up' when transferred back to more 'normal' temperature regimes. It is now generally recognized that the characteristic responses seen in Fig. 15.1 are due primarily to effects on internode extension and, in particular, to effects on cell length (Erwin *et al.*, 1991, 1994), or on cell length and cell division (Strøm and Moe, 1997).

Similar observations to those in Fig. 15.1a were reported by Went (1944) who concluded that growth in length of stems in tomato depended on two sets of processes, one occurring in the light with an optimal temperature near 26°C, and one occurring in darkness with an optimal temperature of 17–19°C. Went coined the term 'thermoperiodicity' to describe these DT–NT effects, and suggested that these were general in higher plants. However, an alternative interpretation of such responses has been put forward by Erwin *et al.* (1989a). These researchers grew lily (*Lilium longiflorum*) plants in 25 combinations of DT and NT and concluded that plant height and internode length were determined by the difference (DIF) between DT and NT rather than by the absolute temperatures used. Temperature combinations in which DT was lower than NT (−DIF) gave short plants, whilst temperature combinations in which DT was higher than NT (+DIF) gave tall plants. Since then, many other research reports have been published which have come to a similar conclusion, and recent reviews on the topic (Erwin and Heins, 1995; Myster and Moe, 1995) suggest that DIF is a general phenomenon in higher plants. Thus, Myster and Moe (1995) give references to 24 species showing strong to medium DIF responses and to only 11 showing little or no response.

In order to clarify whether DT–NT responses are determined by absolute temperatures or by DIF, Langton and Cockshull (1997a) grew plants of tomato and chrysanthemum (*Chrysanthemum morifolium*) for 10 days in 24 combinations of DT and NT, ranging from 12 to 32°C, and recorded the lengths of

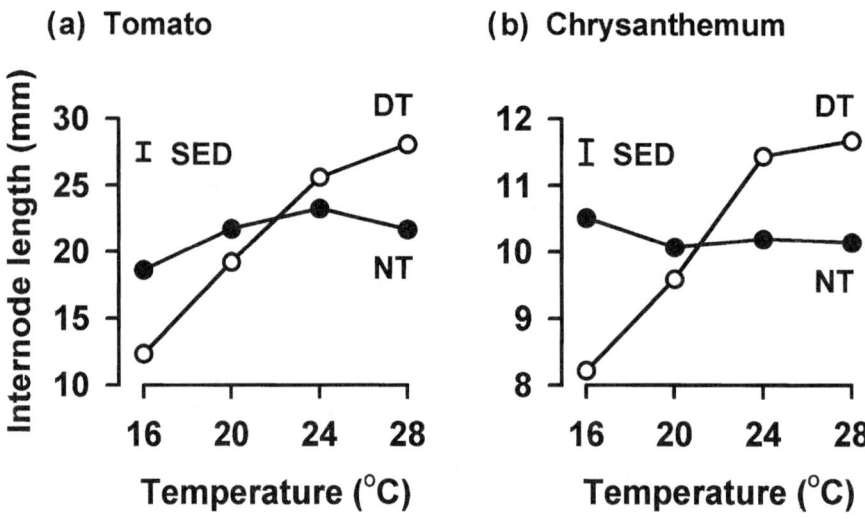

Fig. 15.2. Effects of day temperature (DT) averaged over night temperature and night temperature (NT) averaged over day temperature on internode length (≤1 mm at the start) of (a) tomato and (b) chrysanthemum after 10 days of growth. Bars indicate SED values (6 d.f.) for comparison of DT means and NT means. (Reprinted from Langton, F.A. and Cockshull, K.E. (1997) Is stem extension determined by DIF or by absolute day and night temperatures? *Scientia Horticulturae* 69, 229–237, with kind permission of Elsevier Science, Amsterdam, The Netherlands.)

internodes which had been ≤1 mm at the start of treatment. No relationship could be found between final internode length and DIF for either species and, for a given value of DIF, internode length increased as the absolute DT and NT increased. For example, in combinations giving −4°C DIF in tomato, internode length increased progressively from 4.7 mm at 12°C DT–16°C NT to 24.7 mm at 24°C DT–28°C NT and at 28°C DT–32°C NT.

In contrast, a ready explanation for the responses shown was found by consideration of the absolute temperatures employed. For tomato, factorial analysis of the 4 × 4 (DT × NT) subset of treatments involving the temperatures 16, 20, 24 and 28°C showed clear linear and quadratic effects of both DT and NT (Fig. 15.2a). Internode length increased with an increase in DT up to at least 28°C, which may have been close to the optimum DT for stem extension. An increase in internode length was also given by an increase in NT up to an optimum at around 24°C, but the response to a change in NT was generally much less marked than a response to a change in DT. Interactions between DT and NT proved non-significant, and the second order, quadratic model gave an excellent fit to the data, with 94.7% of variance accounted for. Extending regression to the full data set gave a similarly good fit to the quadratic model, with 93.2% of variance accounted for.

Factorial analysis of the chrysanthemum data also showed there to be a marked linear effect of DT on extension growth, together with a smaller quadratic effect indicating a high DT optimum (Fig. 15.2b). In contrast with tomato, however, a change in NT had relatively little effect on internode length. The second order, quadratic model again gave a good fit to the data, accounting for 89.7% of variance, and extending regression to the full data set gave a similarly good fit, with 87.3% of variance accounted for.

Langton and Cockshull (1997b) further re-appraised published data sets for petunia (*Petunia* × *hybrida*), fuchsia (*Fuchsia* × *hybrida*) and lily, and found that, in all cases, quadratic models based on absolute DT and NT terms gave extremely good fits to the data. Increasing DT gave increased internode extension in all three species, as in tomato and chrysanthemum, with indications of very high DT optima for extension growth. The effects of NT were more varied; in lily and fuchsia, increasing NT decreased internode length, the reverse of the situation in tomato, whilst in petunia, as in chrysanthemum, NT had little or no effect on internode length. Taking all five species together, it was concluded that extension growth responses are determined by absolute DT and NT.

The lack of effect of change in night temperature on internode length in chrysanthemum explains why earlier attempts at Wellesbourne to use temperature 'jumps' during the night to reduce height (by giving a −DIF regime) failed. Erwin and Heins (1995) also reported little or no effect of changes in NT on height in this species. In apparent contrast, Kresten Jensen (1993) found reductions in internode length when a 6°C temperature increase was given to chrysanthemums immediately after sunset, around midnight or before sunrise. However, these increases in NT were accompanied by compensating reductions in DT to preserve the 24-h average temperature, and the extension growth responses could equally well have reflected DT responses rather than NT responses. Although internode lengths were reduced in these treatments, overall plant height frequently was increased due to delayed terminal flower initiation. Similar findings to those for chrysanthemum were reported by Kresten Jensen (1992) for geranium (*Pelargonium* × *hortorum*), but the same objection can be raised that the shorter internodes could equally well have been due to decreased DT.

Cucumber (*Cucumis sativus*) appears to be another species in which stem extension is affected much less by NT than by DT. Thus, Grimstad and Frimanslund (1993) showed that whilst increasing the 24-h temperature by raising DT gave marked increases in internode length, similar increases in the 24-h temperature by raising NT had either no effect or gave a slight reduction in internode length. A lack of effect on stem extension by either increasing or decreasing temperature by 5°C during the first 3 h of the night has been reported for tomato (Gertsson, 1992), and Erwin and Heins (1995) reported that increases in temperature at the beginning or end of the night had no effect on the height of petunia and salvia (*Salvia splendens*). These latter workers found that an increase in temperature during the first 2 h of the night actually increased the height of impatiens (*Impatiens wallerana*), but since NT responses appear to vary widely across species, variation in response to NT manipulation is to be expected.

The effects of DT and NT on tomato and chrysanthemum shown in Fig. 15.2 were obtained with a day length of 12 h (to give day and night equal weight).

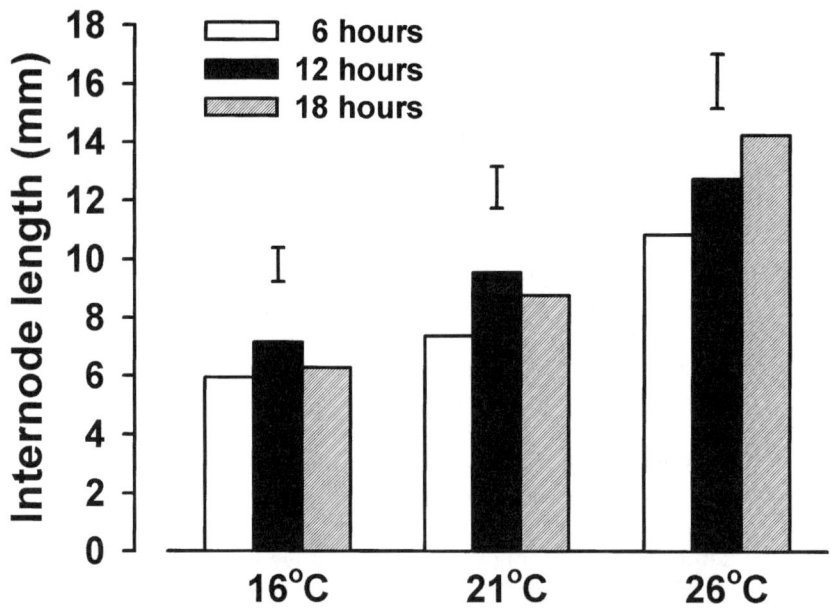

Fig. 15.3. The effects of day length (6, 12 and 18 h) and continuous temperature (16, 21 and 26°C) on internode length (≤1 mm at the start) of chrysanthemum after 10 days of growth. Bars indicate 2 × 95% confidence interval for day lengths within temperatures.

Given that a change in NT appeared to have relatively little effect, the question arose as to whether increasing the length of a high temperature day would increase internode length further, and whether increasing the length of a low temperature day would reduce internode length further. This was tested for both tomato and chrysanthemum by growing plants in the factorial set of treatments given by combining continuous temperatures of 16, 21 and 26°C, with day lengths of 6, 12 and 18 h. In the case of tomato, a change in day length at any given temperature had no significant effect on internode length. Low temperature, on the other hand, markedly depressed internode extension. The results for chrysanthemum (Fig 15.3) were rather similar, but there was some slight evidence that increasing the length of a high temperature day increased internode length. These treatments were applied under conditions of equal daytime irradiance (40 W m^{-2} PAR), and follow-up experiments with irradiances set to give equal daily light integrals (irradiances of 80, 40 and 26.7 W m^{-2} PAR for day lengths of 6, 12 and 18 h respectively) failed to confirm this finding. Overall, it was concluded that increasing the day length beyond 6 hours had no obvious additional effect on internode extension in the two species.

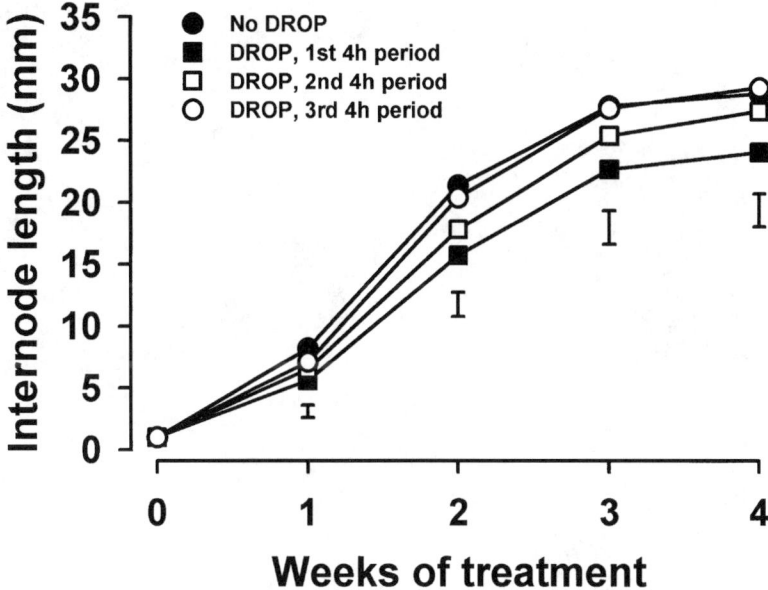

Fig. 15.4. The effect of a temperature reduction (DROP) from 20 to 12°C during the first, second or third 4-h period of a 12-h day on internode extension of tomato. The control treatment was at 20°C continuously. Bars indicate the 5% LSD at each sampling time.

Daytime Sensitivity to Temperature

A probable reason why extending the day length beyond 6 h had little or no additional effect on internode extension in the experiment reported above is that many species appear to show a period of particular sensitivity to temperature in the early part of the day. This was first reported for lily, salvia and petunia by Erwin *et al.* (1989b), who claimed that dropping the temperature during the early part of the day was almost as effective in inhibiting stem elongation as dropping the temperature during the whole of the day. Figure 15.4 shows that tomato is also differentially responsive to short-duration temperature reductions (DROP treatments). The data in Fig. 15.4 are from an experiment in which the temperature was reduced from 20 to 12°C for 4 h, during either the first, second or third 4-h period of a 12-h day, and in which the extension growth of internodes which had been ≤1 mm at the start of treatment was compared with that of similar internodes on plants grown at 20°C continuously. Application of DROP during the first 4 h of the day resulted in significantly shorter internodes at each time of sampling over a 4-week period. Application during the second 4-h period, in the middle of the day, reduced extension growth to a smaller extent. However, DROP applied during the final 4-h period had no significant effect on internode extension.

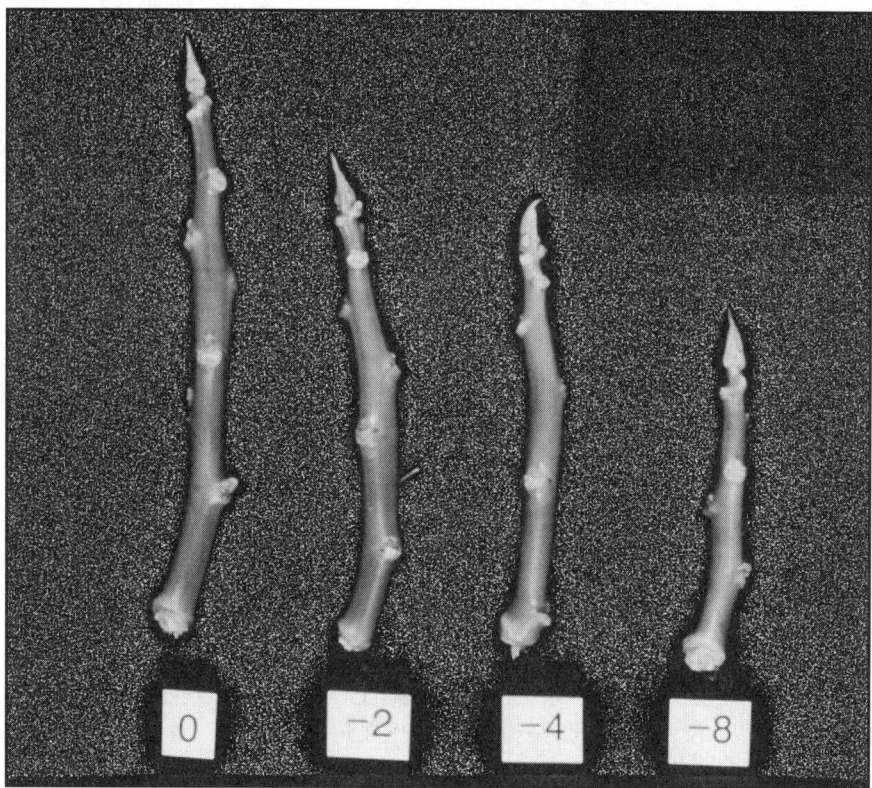

Fig. 15.5. The effects of temperature reductions of 2, 4 and 8°C during the first 6 h of a 12-h day on shoot length of poinsettia. The control (left) was at 18°C continuously. Leaves have been removed for clarity.

Poinsettia is another species which has been shown to respond in this way. Cockshull *et al.* (1995), for example, showed that an 8°C DROP treatment had a much greater effect in reducing internode length when given during the first 6 h of a 12-h day, than when given during the second 6 h, and that increasing the magnitude of the DROP treatment in the first half of the day produced increasingly shorter shoots (Fig. 15.5). A 3-h DROP was less effective than a 6-h DROP, but was still more effective when given during the first 3 h of the day than when given during the second 3 h of the day. Ueber and Hendriks (1992) also found that poinsettia was most sensitive to DROP at the beginning of the day in an experiment where the temperature was reduced from 24 to 8°C during three successive periods starting at dawn. The later the DROP was applied, the less the reduction in extension growth. Rather surprisingly, there was little or no reduction in extension growth when the DROP was from 24 to 16°C. Other species showing the highest sensitivity to temperature at the beginning of the day include chrysanthemum (Cockshull *et al.*, 1995) and fuchsia (Hansen *et al.*, 1996). A rather strange finding reported by Erwin and Heins (1995) was that impatiens was most sensitive to a DT increase when this was given immediately after dawn, but was most sensitive to a DT decrease when this was given in the middle of the day.

In apparent contrast to the above reports, some species (or some species growing in particular environments) have shown no differential sensitivity to temperature during the day. Hansen *et al.* (1996), for example, found that *Verbena* and geranium were equally sensitive to 2-h DROP treatments of 12°C whenever during a 10-h day they were applied. However, these workers found DROP treatments to be ineffective when applied during a 14-h day. Repeatability can also be a problem in experiments to determine sensitivity to DROP treatments, as exemplified by the comparison of trials carried out on four bedding plant species in Sweden (Vogelezang *et al.*, 1992) and the UK (Langton *et al.*, 1992). In Sweden, a 2-h DROP (18°C down to 12°C) applied either immediately before or immediately after dawn reduced plant height compared with the control treatment (no DROP) in salvia, increased plant height in petunia, and had no effect on impatiens or geranium. In contrast, DROP (16°C down to 10°C) applied after dawn in the UK markedly reduced plant height in salvia, petunia and impatiens, and also slightly reduced plant height in geranium, whilst DROP applied before dawn slightly reduced height in salvia and petunia, and had no effect on impatiens and geranium.

In view of these reported differences, it seems hardly surprising that commercial growers have had mixed success in achieving height control by the application of post-dawn DROP treatments. It is at this time during the diurnal cycle that temperature reductions in greenhouses can be most easily obtained, but inconsistency of plant response has, nevertheless, been a general experience. Current research is, therefore, directed towards gaining a better understanding of factors underlying variability of response. In particular, it remains to be seen whether the timing of sensitivity is regulated by a 'light-on' signal (at the start of day) or by an earlier 'light-off' signal (at the end of day). In this latter case, the positioning of the sensitive period, relative to 'dawn', might be expected to vary with night length, and this might be a factor giving rise to variability of response from occasion to occasion. It is not always easy in summer to achieve a significant reduction in DT in commercial growing, and it is tempting to suppose that, if sensitivity is triggered by a 'light-on' signal, it might be possible to achieve a 'day response' during the natural night by associating a reduction in NT with artificial lighting. It is not known, however, how much light is required to signify 'day' from the standpoint of extension growth responses, and this is also currently being investigated.

The Involvement of Gibberellins

Gibberellin (GA) metabolism is almost certainly involved in the regulation of stem extension by temperature. Zieslin and Tsujita (1988) found that GA_3 promoted stem extension in lily growing in a low DT–high NT regime, but not in the reciprocal high DT–low NT regime. This suggests that reduced extension growth in a low DT–high NT regime may be due to the inhibition of GA synthesis. Moe (1990) also observed that the proportional increase in stem

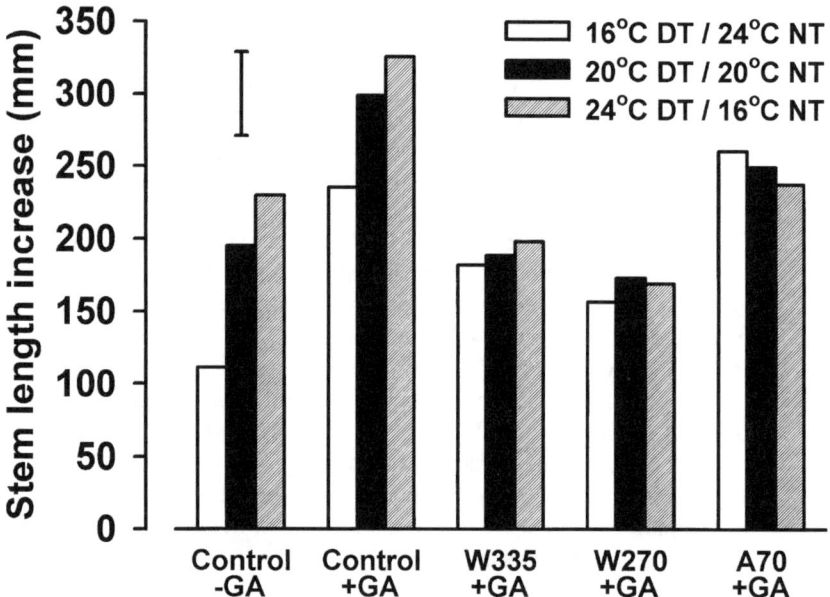

Fig. 15.6. Average stem length increases after 14 days for tomato plants growing in contrasting day–night temperature regimes, following the application of GA_{4+7}. The control plants were cv. Moneymaker, and W335, W270 and A70 were GA-deficient mutants derived from cv. Moneymaker. The bar shows the estimated combined 95% confidence interval. (Reproduced from Langton *et al.*, 1997, by permission of The International Society for Horticultural Science.)

elongation due to GA_3 in *Campanula isophylla* was greater in a low DT–high NT regime than in the reciprocal regime, and observed that the morphogenetic responses of high DT–low NT were mimicked by GA application.

Responses to DT–NT manipulation could involve changes in the concentration of active GAs or, alternatively, changes in the sensitivity of tissues to GA, and studies to resolve this question have been reported by Langton *et al.* (1997). Single drops of GA_{4+7} were applied to the growing tips of three dwarf, single-gene, GA-deficient mutants of tomato and to their wild-type progenitor, and these were grown, together with control plants which had received no GA, in three contrasting DT–NT regimes having the same 24-h average temperature. As can be seen in Fig. 15.6, the wild-type plants without GA showed characteristic height differences after 14 days, reflecting temperature treatments. Wild-type plants which had received GA showed increased extension growth over the controls, particularly those in the low DT–high NT regime, but temperature-mediated differences in height were still apparent at the end of the experiment. The mutant plants which had received GA also showed considerable extension growth, but all three genotypes showed very similar growth in the three different temperature regimes. Since the application of GA to the mutant plants could be expected to have resulted in similar concentrations of GA within the plants, it was reasoned that the similarity of response in the three temperature regimes showed that temperature had not affected plant sensitivity to GA, and that it was likely that temperature affects extension growth in wild-type plants by influencing the concentration of active endogenous GAs.

Essentially similar findings were reported by Maas and van Hattum (1997) working with plants of fuchsia which previously had been treated with the GA synthesis inhibitor, paclobutrazol. When exogenous GAs were applied to such plants, differential growth responses to contrasting DT–NT regimes were no longer shown. We recently have used this approach with wild-type tomato and obtained results very similar to those using mutants described above. However, the essential step in proving (or disproving) a relationship between temperature-mediated stem extension and GA production will be the quantification of concentrations of active GAs within elongating internodes. This has been achieved by Jensen *et al.* (1996) working with *C. isophylla*. These researchers found that temperature regimes stimulating extension growth were accompanied by an increase in the levels of the physiologically active GA_1, and its precursors GA_{19} and GA_{44}, whilst reciprocal temperature regimes depressing extension growth were accompanied by an increased level of GA_{97}, the inactive 2ß-hydroxylated form of GA_{53}, the precursor of GA_{44}. It may be, therefore, that temperature regulates extension growth by interacting with the later stages of the biosynthetic pathway leading to the production of GA_1. Langton *et al.* (1997) also found that the concentrations of the precursors GA_{19}, GA_{20} and GA_{44} were higher in tomato plants grown in a high DT–low NT regime than in those grown in the reciprocal low DT–high NT regime, but unfortunately, recovery levels of the active GA_1 and GA_3 were too low to be reliable.

A Working Hypothesis

Although DIF (the difference between DT and NT) has been widely assumed to regulate temperature-mediated stem extension responses, better and more consistent fits to the published data are given by simple models based on absolute temperatures. Thus, Langton and Cockshull (1997b) conclude that DIF is an artefact lacking real biological significance, and that extension growth is determined by absolute DT and NT. Of these, DT appears to be the dominant factor, and all species so far examined show an increase in extension growth with an increase in DT up to a very high DT optimum.

Response to temperature does not appear to be uniform throughout the day, and a period of particular sensitivity to temperature frequently has been reported. This period generally has been during the early part of the day, and it is likely that it is triggered by a 'light-on' signal. However, it cannot yet be ruled out that sensitivity is triggered by an earlier 'light-off' signal. In either case, light perception is assumed to be an important component of the response, and this may involve either phytochrome or the blue-light photoreceptor. This assumption is based on reports by Moe *et al.* (1991) that the phytochrome photoequilibrium following a low irradiance 'night' (using either incandescent or fluorescent lamps) determined whether differential DT–NT extension growth responses were shown in *Campanula*, and by Maas and van Hattum (1997) that blue light during the day was essential for differential DT–NT extension growth responses in fuchsia. It is hypothesized that the 'light-on' or 'light-off' signal triggers GA production or interconversion since there is evidence, for example, that phytochrome mediates direct changes in GA metabolism affecting stem elongation in cowpea, *Vigna sinensis* (Fang *et al.*, 1991). It is hypothesized further that DT interacts with these GA responses to affect the concentration of active GAs, probably GA_1, translocated to elongating internodes. This working hypothesis underlies current studies at HRI on the regulation of stem extension by temperature.

References

Cockshull, K.E., Langton, F.A. and Cave, C.R.J. (1995) Differential effects of different DIF treatments on chrysanthemum and poinsettia. *Acta Horticulturae* 378, 15–25.

Erwin, J.E. and Heins, R.D. (1995) Thermomorphogenic responses in stem and leaf development. *HortScience* 30, 940–949.

Erwin, J.E., Heins, R.D. and Karlsson, M.G. (1989a) Thermomorphogenesis in *Lilium longiflorum*. *American Journal of Botany* 76, 47–52.

Erwin, J.E., Heins, R.D., Berghage, R.D., Kovanda, B.J., Carlson, W.H. and Biernbaum, J.A. (1989b) Cool mornings can control plant height. *GrowerTalks* 52, 73–74.

Erwin, J.E., Velguth, P. and Heins, R.D. (1991) Diurnal variations in temperature affect cellular elongation but not division. *HortScience* 26, 105.

Erwin, J.E., Velguth, P. and Heins, R.D. (1994) Diurnal temperature fluctuations affect *Lillium* cell elongation but not division. *Journal of Experimental Botany* 45, 1019–1025.

Fang, N., Bonner, B.A. and Rappaport, L. (1991) Phytochrome mediation of gibberellin metabolism and epicotyl elongation in cowpea, *Vigna sinensis* L. In: Takahashi, N., Phinney, B.O. and MacMillan, J. (eds), *Gibberellins*. Springer-Verlag, New York, pp. 280–288.

Gertsson, U.E. (1992) Influence of temperature on shoot elongation in young tomato plants. *Acta Horticulturae* 327, 71–76.

Grimstad, S.O. and Frimanslund, E. (1993) Effect of different day and night temperature regimes on greenhouse cucumber young plant production, flower bud formation and early yield. *Scientia Horticulturae* 53, 191–204.

Hansen, H.T., Hendriks, L., Ueber, E. and Andersen, A.S. (1996) Effect of a low temperature period (drop) during different periods on morphogenesis of *Dendranthema* 'Surf', *Fuchsia* 'Beacon', *Verbena* 'Karminrosa', and *Pelargonium* 'Pulsar Red'. *Gartenbauwissenschaft* 4 96, 150–158.

Heuvelink, E. (1989) Influence of day and night temperature on the growth of young tomato plants. *Scientia Horticulturae* 38, 11-22.

Jensen, E., Eilertsen, S., Ernsten, A., Juntilla, O. and Moe, R. (1996) Thermoperiodic control of stem elongation and endogenous gibberellins. *Journal of Plant Growth Regulation* 15, 167-171.

Kresten Jensen, H.E. (1992) Effects of duration and placing of the high night DIF temperature on the morphogenesis of *Pelargonium* × *zonale* hybrid 'Pink Cloud'. *Acta Horticulturae* 327, 17–23.

Kresten Jensen, H.E. (1993) Influence of duration and placement of a high night temperature on morphogenesis of *Dendranthema grandiflora* Tzvelev. *Scientia Horticulturae* 54, 327-335.

Langton, F.A. and Cockshull, K.E. (1997a) Is stem extension determined by DIF or by absolute day and night temperatures? *Scientia Horticulturae* 69, 229-237.

Langton, F.A. and Cockshull, K.E. (1997b) A re-appraisal of DIF extension growth responses. *Acta Horticulturae* 435, 57-64.

Langton, F.A., Cockshull, K.E., Cave, C.R.J. and Hemming, E.J. (1992) Temperature regimes to control plant stature: current UK R&D. *Acta Horticulturae* 327, 49-59.

Langton, F.A., Lumsden, P.J. and Horridge, J. (1997) Are gibberellins involved in temperature-mediated stem extension responses in tomato? *Acta Horticulturae* 435, 105-112.

Maas, F.M. and van Hattum, J. (1997). The role of gibberellins in the thermo- and photocontrol of stem elongation in *Fuchsia*. *Acta Horticulturae* 435, 93-104.

Moe, R. (1990) Effect of day and night temperature alternations and of plant growth regulators on stem elongation and flowering of the long-day plant *Campanula isophylla* Moretti. *Scientia Horticulturae* 43, 291-305.

Moe, R., Heins, R.D. and Erwin, J.E. (1991) Effect of day and night temperature alternations, and plant growth regulators on stem elongation and flowering of the long-day plant *Campanula isophylla* Moretti. *Scientia Horticulturae* 48, 141-151.

Myster, J. and Moe, R. (1995) Effect of diurnal temperature alternations on plant morphology in some greenhouse crops - a mini review. *Scientia Horticulturae* 62, 205-215.

Strøm, M. and Moe, R. (1997) DIF affects internode and cell extension growth and cell number in *Campanula isophylla* shoots. *Acta Horticulturae* 435, 17-24.

Ueber, E. and Hendriks, L. (1992) Effects of intensity, duration and timing of a temperature drop on the growth and flowering of *Euphorbia pulcherrima* Willd. ex Klotzsch. *Acta Horticulturae* 327, 33-40.

Vogelezang, J.V.M., Moe, R., Schüssler, H.K., Hendriks, L., Cuijpers, L.H.M. and Ueber, E. (1992) Cooperative European research on temperature strategies for bedding plants. *Acta Horticulturae* 327, 11-16.

Went, F.W. (1944) Plant growth under controlled conditions. II. Thermoperiodicity in growth and fruiting of the tomato. *American Journal of Botany* 31, 135-150.

Zieslin, N. and Tsujita, M.J. (1988) Regulation of stem extension of lilies by temperature and the effect of gibberellin. *Scientia Horticulturae* 37, 165-169.

16 Modification of Plant Morphology by Genetic Manipulation of Gibberellin Biosynthesis

P. Hedden[1], J.P. Coles[1], A.L. Phillips[1], S.G. Thomas[1], D.A. Ward[1], I.S. Curtis[2], J.B. Power[2], K.C. Lowe[2] and M.R. Davey[2]

[1]*IACR-Long Ashton Research Station, Department of Agricultural Sciences, University of Bristol, Long Ashton, Bristol BS41 9AF, UK; and [2]Plant Science Division, School of Biological Sciences, University of Nottingham, University Park, Nottingham NG7 2RD, UK*

Introduction

Chemical modification of plant morphology is common agronomic practice (Hedden and Hoad, 1994; Gianfagna, 1995). In many cases, such treatments serve to manipulate the content of the gibberellin (GA) hormones in plant tissues. For example, GA_3 (gibberellic acid) is used to stimulate berry growth in seedless grapes, while mixtures of GA_4 and GA_7 are used to induce reproductive development in conifers and to enhance fruit quality in apple. Conversely, plant stature is controlled in many species, particularly cereals, top fruit and ornamentals, by application of growth retardants, most of which act by inhibiting GA biosynthesis (Hedden and Hoad, 1994). However, the long-term usage of such chemicals may have detrimental effects on the environment. We are pursing an alternative, non-chemical approach to the control of plant development, in which GA content is modified by genetic manipulation of the biosynthetic pathway. Although mutants with reduced rates of GA biosynthesis are known for many species, and some, such as the semi-dwarf pea varieties containing the *le* mutation, are used commercially, suitable mutations are not yet available in most of the important crop species.

GAs function as endogenous regulators of plant growth, controlling development throughout the life cycle of higher plants (Hooley, 1994). Thus, they influence such processes as seed germination, stem and root extension, leaf expansion, flower development and fruit growth. They also act as mediators of environmental stimuli, as in the bolting response of rosette plants after exposure to long days or low (vernalizing) temperatures. Since the concentration of endogenous GAs limits the rate and extent of plant growth in many cases, regulation of GA biosynthesis is an important factor in the control of development. Identifying the biosynthetic steps that are subject to developmental or environmental regulation is necessary for elucidating the underlying mechanisms in GA-mediated control of plant growth. The activity of enzymes that catalyse such steps is likely to have a large influence on the concentration of the active GAs and, consequently, these enzymes form prime targets for the genetic manipulation of GA biosynthesis.

We are using two plant species with different growth habits as model systems to examine the effects on development of genetic manipulation of GA biosynthesis and to determine the feasibility of this approach for producing plants with desirable morphology. *Arabidopsis thaliana* is an annual rosette plant in which bolting is stimulated by exposure to long days and vernalizing temperatures, although there is normally not an absolute requirement for these conditions. Bolting in *Arabidopsis* is dependent on GA production. The species is extremely valuable as an experimental system because it is easily transformed and has a short life cycle. We are also using *Solanum dulcamara*, a perennial plant with a caulescent growth habit which is easy to transform (Curtis *et al.*, 1998) and can be propagated rapidly through stem cuttings. The choice of target genes for genetic manipulation of these species is discussed in the following section.

Gibberellin Biosynthesis

The GAs constitute a very large group of diterpenoid tetracyclic carboxylic acids, with currently 121 different structures characterized from higher plants and fungi. However, relatively few of these structures have intrinsic biological activity, with many of the other compounds being biosynthetic precursors, metabolites or inactive by-products produced in the biosynthesis of the active forms. The predominant biologically active GAs controlling stem extension are GA_1 and GA_4, which differ in the presence or absence, respectively, of a hydroxyl group on C-13. The biosynthetic pathway to these GAs and their 2ß-hydroxylated (inactive) metabolites is outlined in Fig. 16.1. GA biosynthesis has been discussed thoroughly in recent reviews (Hedden and Kamiya, 1997; MacMillan, 1997) and will be described only briefly here. As diterpenes, GAs are formed from geranylgeranyl diphosphate, which is cyclized in two steps, first to copalyl diphosphate and then to *ent*-kaurene, a tetracyclic hydrocarbon. These reactions, which occur in proplastids (Aach *et al.*, 1995), are catalysed by copalyl diphosphate synthase (CPS) and *ent*-kaurene synthase (EKS), respectively. cDNA clones encoding these enzymes have been isolated (Sun and Kamiya, 1994; Bensen *et al.*, 1995; Ait-Ali *et al.*, 1996; Yamaguchi *et al.*, 1996); their derived amino acid sequences contain a non-polar N terminus that may function to target the enzymes to the plastid. In the case of CPS, but not EKS, it was possible to demonstrate import into plastids, during which the targeting sequence was cleaved (Sun and Kamiya, 1994).

ent-Kaurene is oxidized on membranes by cytochrome P_{450}-dependent mono-oxygenases, resulting in the formation of GA_{12} 7-aldehyde, the first intermediate with the *ent*-gibberellane skeleton. Oxidation of the 7-aldehyde to a carboxylic acid yields GA_{12}, which is commonly hydroxylated on C-13 to give

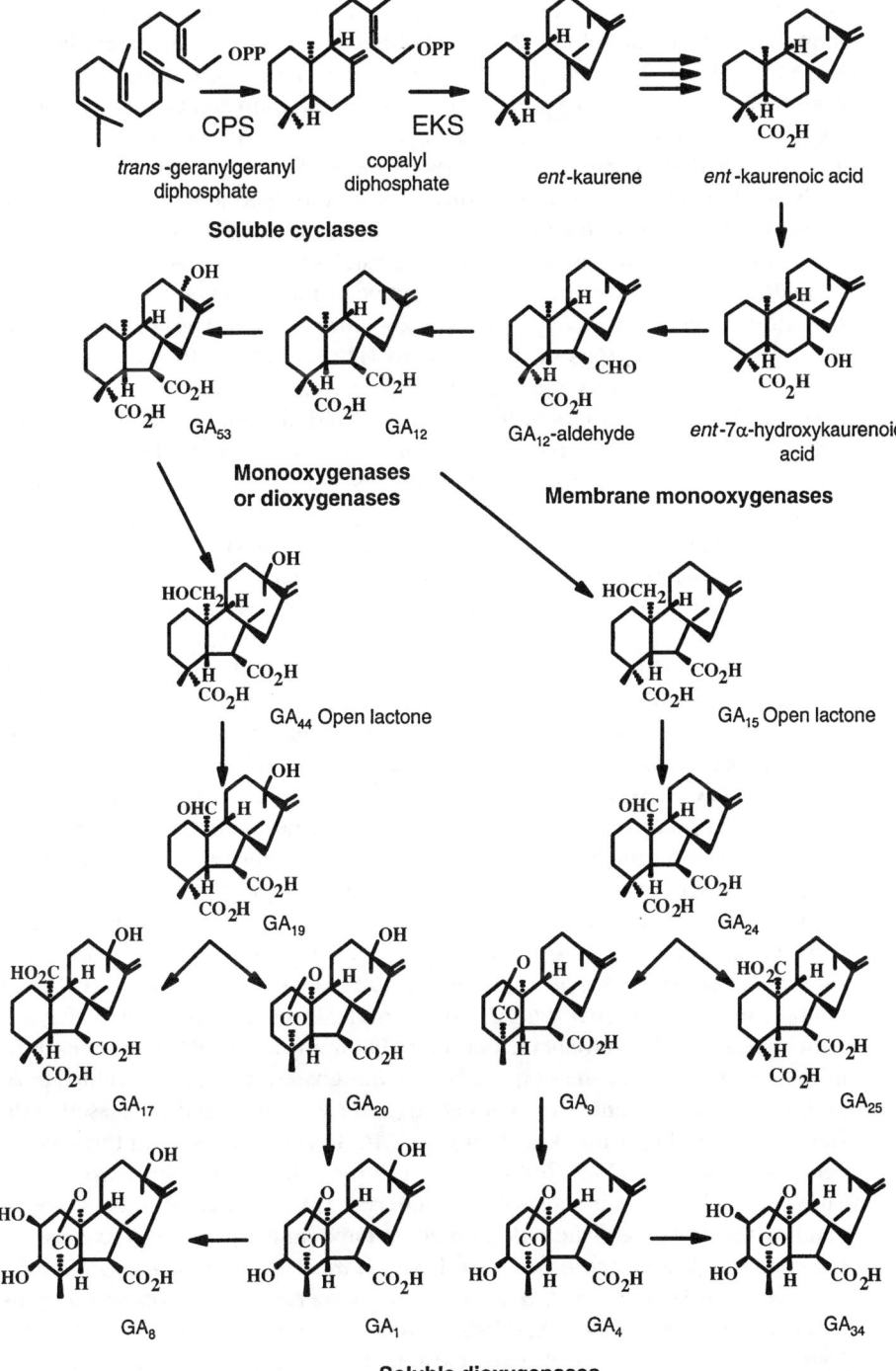

Fig. 16.1. Biosynthetic pathways to the biologically active GA_1 and GA_4 and their inactive 2ß-hydroxy metabolites, GA_8 and GA_{34}. The tricarboxylic acids GA_{17} and GA_{25} are biologically inactive by-products of GA 20-oxidase activity.

GA_{53}. The 7-oxidation and 13-hydroxylation reactions may be catalysed either by mono-oxygenases or soluble 2-oxoglutarate-dependent dioxygenases, depending on the tissue. For example, in pumpkin (*Cucurbita maxima*) endosperm, both types of 7-oxidase are present and 13-hydroxylation is catalysed by a monooxygenase (Hedden et al., 1984), whereas a dioxygenase catalyses 13-hydroxylation in spinach (*Spinacea oleracea*) leaves (Gilmour et al., 1986). Carbon-20 in GA_{12} and GA_{53} is oxidized to the corresponding aldehydes, from which C-20 is cleaved to give the C_{19} GA products, GA_9 and GA_{20}, respectively, the complete sequence of reactions being catalysed by GA 20-oxidases. Finally, 3ß-hydroxylation yields the biologically active products, GA_4 and GA_1, which are then turned over to inactive products by 2ß-hydroxylation. The enzymes responsible for these later steps of the pathway, GA 20-oxidase, 3ß-hydroxylase and 2ß-hydroxylase, are dioxygenases that utilize 2-oxoglutarate as a co-substrate, have an absolute requirement for Fe^{2+} and are stimulated by ascorbate. Oxidative decarboxylation of 2-oxoglutarate to succinate results in the formation at the active site of a ferryl ($Fe^{4+}=O$) species that oxidizes the GA substrate (Hedden, 1997).

Gibberellin 20-oxidases are multifunctional enzymes that catalyse each step in the sequential oxidation of C-20. The first GA 20-oxidase cDNA to be cloned was derived from immature cotyledons of pumpkin (Lange et al., 1994). The enzyme encoded by this cDNA is unusual in that it converts GA_{12} into GA_{25} by oxidizing the C-20 methyl group to a carboxylic acid group; the aldehyde intermediate in this reaction sequence (GA_{24}) is also converted by the enzyme into the C_{19} product, GA_9, but only in about 1% yield (Lange et al., 1994). GA_{53} is oxidized by the enzyme in a similar manner, but with lower efficiency than GA_{12}. At least 20 GA 20-oxidase cDNAs have now been cloned but, in contrast to the enzyme from pumpkin seed, all other examples catalyse the removal of C-20 as the major reaction and form the tricarboxylic acids as minor (< 5%) by-products. The phylogenetic relationship of the GA 20-oxidases, produced on the basis of a highly conserved region of their amino acid sequence, is shown in Fig. 16.2. In general, the sequences are grouped according to plant families. It has been shown for several species that the GA 20-oxidases are encoded by small multigene families. In *Arabidopsis*, for example, three genes encoding enzymes with similar function have been identified (Phillips et al., 1995). The genes differ in their patterns of expression, with one, designated *At2301* by Phillips et al. (1995) being equivalent to the *GA5* gene (Xu et al., 1995), and expressed in the stem of the bolting plant. A second gene (*At2353*) is expressed in the inflorescence and the third (*YAP169*) in the developing silique (Phillips et al., 1995). These genes are, thus, associated with different GA-regulated processes. Within families, the most highly homologous genes have similar patterns of expression. For example, *pea2* (*Pisum sativum*; Lester et al., 1996) and *bean3* (*Phaseolus vulgaris*; Garcia-Martinez et al., 1997) are expressed in developing embryos, whereas *pea1* (Martin et al., 1996; Garcia-Martinez et al., 1997) and *bean1* (Garcia-Martinez et al., 1997) are expressed in vegetative tissues and young seeds.

Many plant species, including *Arabidopsis* (Talon et al., 1990), accumulate C_{20} GAs in the shoots, indicating that the conversion of C_{20} into C_{19} GAs,

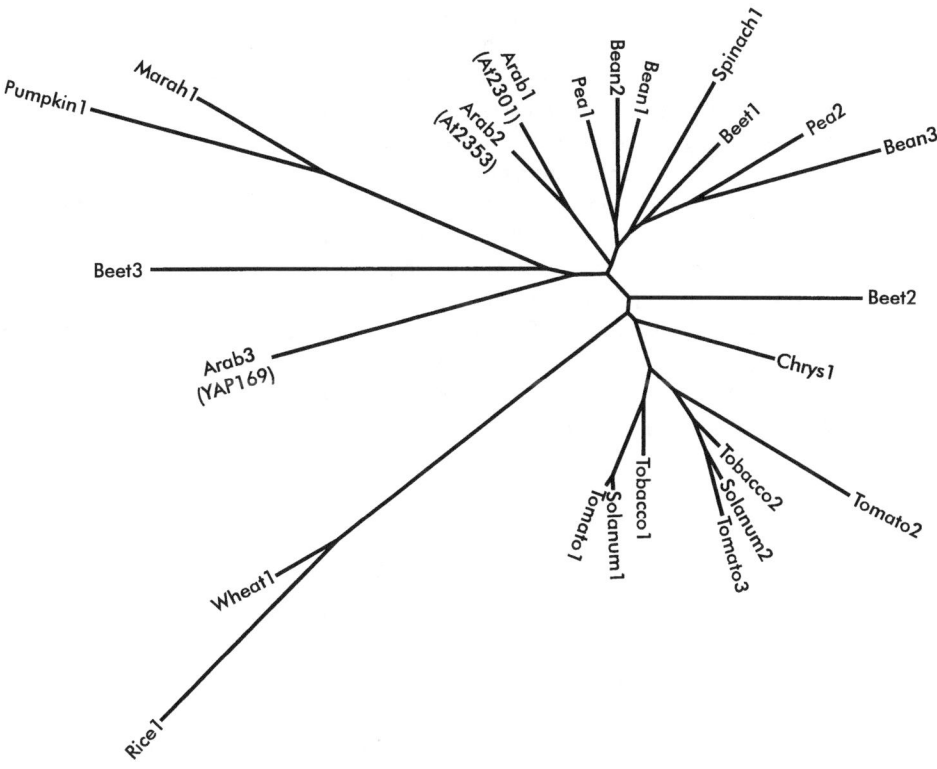

Fig. 16.2. Unrooted tree showing the phylogenetic relationship of GA 20-oxidases. Sequence comparisons were made over 75 amino acids using CLUSTAL X, and the phylogenetic relationships were determined using the PHYLIP programs PROTDIST and NEIGHBOR. Arab, *Arabidopsis thaliana*; bean, *Phaseolus vulgaris*; beet (sugar beet), *Beta vulgaris*; chrys, *Chrysanthemum morifolium*; Marah, *Marah macrocarpus*; pumpkin, *Cucurbita maxima*; Solanum, *Solanum dulcamara*; tobacco, *Nicotiana affinis*.

catalysed by GA 20-oxidase, is a relatively slow step in GA biosynthesis. This is consistent with the function of GA 20-oxidases as regulatory enzymes. As well as being under developmental control, expression of GA 20-oxidase genes is negatively regulated by the action of GA (Phillips *et al.*, 1995; Xu *et al.*, 1995; Toyomasu *et al.*, 1997). This feedback regulation, which has been found also for GA 3ß-hydroxylase (Chiang *et al.*, 1995), provides a mechanism by which plants maintain GA homeostasis. A third regulatory factor has been demonstrated in long-day plants, in which GA 20-oxidase gene expression is controlled by photoperiod (Wu *et al.*, 1996) and is a key determinant in the bolting response (Gilmour *et al.*, 1986).

Being aware of their regulatory, potentially rate-determining roles in GA biosynthesis, we are investigating GA 20-oxidases as targets for the genetic manipulation of GA biosynthesis. As described below, we have altered the expression of each of the GA 20-oxidase genes in *Arabidopsis*, primarily to test the utility of this approach to modifying plant development. The work should provide basic information on the importance of these genes in determining the concentration of bioactive GAs, as well as on the developmental processes with which the individual GA 20-oxidase genes are involved. Thus, we have overexpressed the GA 20-oxidase genes and also attempted to reduce the expression of each gene by introducing antisense sequences. In addition, in an alternative strategy to reduce endogenous GA content, we have overexpressed the pumpkin GA 20-oxidase gene. Since the pumpkin enzyme converts the aldehyde intermediates predominantly to the biologically inactive carboxylic acids, the presence of large amounts of this enzyme should divert the pathway, so reducing the amount of the aldehydes available to the endogenous 20-oxidase. We anticipated that the result should be a reduction in amounts of C_{19} GAs produced. Results from these experiments are summarized below. Full details will be published elsewhere.

Transformation of *Arabidopsis thaliana*

The three *Arabidopsis* GA 20-oxidase genes (*At2301/GA5*, *At2353* and *YAP169*) were overexpressed in *Arabidopsis* cv. Columbia. The full-length coding sequences behind a double cauliflower mosaic virus (CaMV) 35S promoter in a binary vector containing the *NPTII* gene were introduced into *Arabidopsis* via *Agrobacterium tumefaciens*-mediated transformation, using vacuum infiltration to infect the plants (Bechtold *et al.*, 1993). Transformants, which were selected by germinating seeds on kanamycin, bolted earlier than wild-type plants or plants that had been transformed with the binary vector lacking the GA 20-oxidase insert (controls). In the progeny from the GA 20-oxidase transformants, early bolting appeared to segregate with gene dosage. Plants that were homozygous for the transgene bolted earlier than hemizygous plants, which, in turn, bolted before azygous lines (Fig. 16.3a). Progeny from the homozygous lines were used for detailed characterization. Since the three *Arabidopsis* GA 20-oxidases have similar functions and were behind a strong, constitutive promoter, overexpression of each gene predictably produced plants that were phenotypically identical. The effects were apparent in the seedlings, which possessed longer hypocotyls and petioles than the controls and untransformed cv. Columbia, resembling controls that had been treated with GA_3. As well as flowering earlier, the 20-oxidase-overexpressing lines produced stems that were substantially taller than the controls at maturity (Fig. 16.3b), these effects being more pronounced when the plants were grown in short days than in long days. Analysis of the GA content in the shoots of the bolting plants confirmed that the concentration of C_{19} GAs was elevated in the plants overexpressing GA 20-oxidase cDNAs. Furthermore, the contents of the aldehyde intermediates were reduced, consistent with a higher rate of turnover of these precursors. These observations indicate that 20-oxidase activity limits the steady-state concentration of active GAs and is an important determinant of growth rate throughout the life cycle of the *Arabidopsis* plant.

Seedlings from lines expressing antisense mRNA for the stem-specific GA

Fig. 16.3. Transgenic *Arabidopsis* plants with altered expression of GA 20-oxidase genes. (a) Segregation of the *35S::YAP169* gene in second generation lines. The tray contains homozygous (2); hemizygous (1), azygous (0) and untransformed (W) cv. Columbia plants. (b) Comparison of plants overexpressing GA 20-oxidase (*At2301*) or expressing antisense *At2301* mRNA with a control line. Plants were grown in short days (10-h photoperiod).

20-oxidase gene (*GA5/At2301*) developed shorter hypocotyls and petioles than controls and the wild-type. In long days, flowering occurred at the same time as for the controls, but there was a reduction in stem length; in short days, flowering was delayed and there was a severe reduction in stem height (Fig. 16.3b), this reduction being much more severe than that in long days. Treatment with GA$_3$ fully restored the growth of the seedlings and the flowering plants to that of the control, indicating that only GA biosynthesis had been modified in these plants. There was no effect of the antisense expression of *At2301* on the numbers and fertility of the flowers, or on growth of the siliques. Northern analysis showed that, in the *At2301* antisense lines, transcript abundance for the native gene in the upper shoot was extremely low. It is possible, however, that GA production via the *At2353* GA 20-oxidase, the transcript of which accumulates in the infloresence and is not down-regulated in the *At2301* antisense lines, may partially compensate for loss of *At2301* GA 20-oxidase activity. Since flower production is initiated in *Arabidopsis* before stem elongation begins, GAs from the reproductive tissues may contribute to stem extension. The *GA5* mutant, which essentially contains a null mutation in the stem-specific GA 20-oxidase gene (*At2301*) (Xu *et al.*, 1995), produces some stem elongation and contains 10-30% of the C$_{19}$ GAs in the shoot relative to the Landsberg *erecta* wild-type (Talon *et al.*, 1990), confirming that other GA 20-oxidases can provide GAs to the shoot. In one transgenic line that, on the basis of a Northern blot, expresses antisense mRNA for the *At2353* gene, growth of the seedling and of older plants in long days was normal, whereas in short days, stem height and flower fertility were reduced. The mildness of the phenotype in this case is probably due to only a small reduction in expression of the native gene. Plants have not yet been obtained in which expression of an antisense *YAP169* gene persisted into the second generation. However, one line, in which antisense *YAP169* mRNA was expressed in the first generation, produced short hypocotyls and petioles at the early seedling stage, but grew normally thereafter. It is possible, therefore, that *YAP169*, which we suspect may be expressed in developing seeds, is involved in early seedling growth and, possibly, also in seed germination.

After transformation with the pumpkin GA 20-oxidase cDNA, a range of different phenotypes were obtained. In one line, stem height was slightly greater than that in the control, while in others it was slightly reduced. The taller line flowered earlier than the controls and was, therefore, similar to the lines which overexpressed the *Arabidopsis* GA 20-oxidase genes. The upper shoot of this line contained an elevated content of GA$_4$, consistent with its higher growth rate. We found that transcript abundance for the native *At2301/GA5* and *At2353* genes in the upper stems of these plants was similar to that in the controls, despite a report that mRNA levels for *GA5* were elevated in plants overexpressing the pumpkin GA 20-oxidase gene (Xu and Zeevaart, 1997). These workers found little effect of the transgene on stem height and suggested that up-regulation of the endogenous GA 20-oxidase gene maintained the concentration of active GAs. They did, however, find a very large increase in the content of the tricarboxylic acid, GA$_{17}$, which is consistent with the presence of the pumpkin enzyme in high abundance.

Fig. 16.4. A transgenic plant of *S. dulcamara* expressing the pumpkin GA 20-oxidase gene (right) and an untransformed plant (left) shortly after transplanting from culture to soil.

Transformation of *Solanum dulcamara*

Solanum dulcamara (bittersweet or woody nightshade) is grown commercially in its variegated form as an ornamental plant. Since it is very easy to transform and can be propagated rapidly from cuttings, we have been using *S. dulcamara* as a model system in our work to control the growth of ornamental plants by genetic manipulation of GA biosynthesis. We have used *A. tumefaciens* strain GV3101 for transformation of leaf discs to introduce the pumpkin GA 20-oxidase cDNA behind the CaMV 35S promoter. In contrast to the results obtained with *Arabidopsis*, each of several independent *S. dulcamara* lines expressing the pumpkin gene had the same phenotype. The plants were dwarfed in culture but, on transfer to soil, grew more vigorously, although stem height remained less than that of untransformed plants (Fig. 16.4). The transgenic plants produced intense green leaves, containing more chlorophyll and carotenoids than the wild-type, and the stems were also more deeply pigmented. Flower formation, pollen viability and fruit set were normal in the transgenic lines. Comparison between the GA content in shoots of one such line and the wild-type revealed that the concentration of GA_1, the major biologically active GA in *S. dulcamara*, was reduced in all tissues, particularly in the upper stem, in which it is about 10% the concentration of that present in wild-type stems. The concentrations of all other members of the early-13-hydroxylation pathway (Fig. 16.1) were also reduced, except for that of the tricarboxylic acid GA_{17}, which, while hardly detectable in wild-type tissues, was the most abundant 13-hydroxy GA in the pumpkin 20-oxidase-expressing line. These findings confirmed that the transgene was being expressed and that its enzyme product was active. Unexpectedly, however, the concentration of the non-13-hydroxylated GA, GA_4, was higher in the transgenic line than in wild-type plants, an observation that may explain why there was only a modest reduction in stem height in the former. It is possible that high amounts of activity of the pumpkin GA 20-oxidase may increase the flux through the non-13-hydroxylation pathway which, in combination with enhanced expression of the native GA 20-oxidase gene, could result in greater production of GA_4. We currently are seeking evidence for this proposal by analysing the concentration of all members of the non-13-hydroxylated GA pathway and comparing the transcript abundance for the native GA 20-oxidase genes in the transgenic and non-transformed lines.

As part of the same project, we have used PCR to obtain DNA fragments that, on the basis of their sequences, we believe to be derived from two different *S. dulcamara* GA 20-oxidase genes. We recently have transformed *S. dulcamara* with antisense copies of these sequences behind the 35S promoter and will be able to determine if silencing either of these genes will give the desired compact growth habit.

Conclusions

We have demonstrated that modification of GA 20-oxidase gene expression in *Arabidopsis* results in developmental changes. Overexpression causes enhanced seedling growth, accelerated flowering and increased stem elongation. This confirms the importance of GA 20-oxidase in determining the content of biologically active GAs and, thereby, in the control of GA-mediated development. Antisense expression should provide a means to determine the function of the three *Arabidopsis* GA 20-oxidase genes, which, on the basis of their distinct tissue-specific patterns of expression (Phillips *et al.*, 1995), are presumed to be involved in different developmental processes. We have achieved a severe reduction in expression of the stem-specific gene (*GA5*), as determined by the amount of transcript produced. Silencing of this gene reduces seedling and stem growth, and, in short days, delays flowering; the effect on flowering time and stem length is more pronounced in short days than in long days, consistent with a greater sensitivity to GAs in long days (Reed *et al.*, 1996; Xu *et al.*, 1997). We have, so far, obtained only a modest reduction in expression of the antisense mRNA for the inflorescence-specific gene (*At2353*) and have no plants in which expression of antisense *YAP169* (silique-specific gene) survived into the second generation. The altered phenotype for the antisense *At2353* plant was apparent only in short days, when it resembled a recently described dwarf mutant (*ga6*) (Sponsel *et al.*, 1997). The *ga6* mutant is thought, like *ga5*, to be defective in GA 20-oxidase activity and may have a lesion in the *At2353* gene. It would appear that stem height in *Arabidopsis* is controlled by at least two GA 20-oxidase genes.

Although ectopic expression of the pumpkin GA 20-oxidase cDNA provides, in theory, an attractive strategy for reducing GA content by diverting intermediates to inactive by-products, the results of this approach have been variable. In *Arabidopsis*, we obtained both overgrowth and reduced-growth phenotypes, while in *S. dulcamara*, we obtained consistently reduced growth, although the degree of dwarfing varied depending on the developmental stage. Several factors, including the relative levels of expression of the transgene and the endogenous GA 20-oxidase genes, and the availability of substrates for the gene products, are likely to influence the outcome of expressing the pumpkin GA 20-oxidase gene. Each of these factors will vary with tissue and developmental stage. The utility of this approach, therefore, is likely to vary according to species and, in view of the complexity of the system, the outcome will be largely empirical. In contrast, our work with expression of antisense GA 20-oxidase genes has confirmed the feasibility of this approach to restrict growth. Furthermore, our work is producing fundamental information on the mechanisms by which plants maintain GA concentrations and on the contributions of individual biosynthetic enzymes to this process.

Acknowledgements

We are grateful to Stephen Croker, Simon Magnes and Raúl García-Lepe for gibberellin analyses. This work was supported by the Biotechnology and Biological Sciences Research Council and the Ministry of Agriculture, Fisheries and Food of the United Kingdom under Open Contract No. 9407.

References

Aach, H., Böse, G. and Graebe, J.E. (1995) *ent*-Kaurene biosynthesis in a cell-free system from wheat (*Triticum aestivum* L.) seedlings and the localization of *ent*-kaurene synthetase in plastids of three species. *Planta* 197, 333-342.

Ait-Ali, T., Swain, S.M., Reid, J.B., Sun, T.P. and Kamiya, Y. (1996) The *Ls* locus of pea encodes the gibberellin biosynthesis enzyme *ent*-kaurene synthase A. *The Plant Journal* 11, 443-454.

Bechtold, N., Ellis, J. and Pelletier, G (1993) *Agrobacterium*-mediated gene-transfer by infiltration of adult *Arabidopsis thaliana* plants. *Comptes Rendus de l'Academie des Sciences Serie III - Sciences de la Vie* 316, 1194-1199.

Bensen, R.J., Johal, G.S., Crane, V.C., Tossberg, J.T., Schnable, P.S., Meeley, R.B. and Briggs, S.P. (1995) Cloning and characterization of the maize *An1* gene. *The Plant Cell* 7, 75-84.

Chiang, H.H., Hwang, I. and Goodman, H.M. (1995) Isolation of the *Arabidopsis GA4* locus. *The Plant Cell* 7, 195-201.

Curtis, I.S., Power, J.B., Hedden, P., Phillips, A., Ward, D., Lowe, K.C. and Davey, M.R. (1998) Production of transgenic plants of *Solanum dulcamara* L. - an important semi-woody plant. In: Davey, M.R., Alderson, P.G., Lowe, K.C. and Power, J.B. (eds), *Tree Biotechnology: Towards the Millenium*. University of Nottingham Press, Nottingham.

García-Martínez, J.L., López-Diaz, I., Sánchez-Beltrán, M.J., Phillips, A.L., Ward, D.A., Gaskin, P. and Hedden, P. (1997) Isolation and transcript analysis of gibberellin

20-oxidase genes in pea and bean in relation to fruit development. *Plant Molecular Biology* 33, 1073-1084.

Gianfagna, T. (1995) Natural and synthetic growth regulators and their use in horticultural and agronomic crops. In: Davies, P.J (ed.), *Plant Hormones: Physiology, Biochemistry and Molecular Biology*, 2nd Edn. Kluwer Academic Publishers, Dordrecht, pp. 751-773.

Gilmour, S.J., Zeevaart, J.A.D, Schwenen, L. and Graebe, J.E. (1986) Gibberellin metabolism in cell-free extracts from spinach leaves in relation to photoperiod. *Plant Physiology* 82, 190-195.

Hedden, P. (1997) The oxidases of gibberellin biosynthesis: their function and mechanism. *Physiologia Plantarum* 101, 709-719.

Hedden, P. and Hoad, G.V. (1994) Growth regulators and crop productivity. In: Basra, A.S. (ed.) *Mechanisms of Plant Growth and Improved Productivity: Modern Approaches*. Marcel Dekker, New York, pp. 173-198.

Hedden, P. and Kamiya, Y. (1997) Gibberellin biosynthesis: enzymes, genes and their regulation. *Annual Review of Plant Physiology and Plant Molecular Biology* 48, 431-460.

Hedden, P., Graebe, J.E., Beale, M.H., Gaskin, P. and MacMillan, J. (1984) The biosynthesis of 12α-hydroxylated gibberellins in a cell-free system from *Cucurbita maxima* endosperm. *Phytochemistry* 23, 569-574.

Hooley, R. (1994) Gibberellins: perception, transduction and responses. *Plant Molecular Biology* 26, 1529-1555.

Lange, T., Hedden, P. and Graebe, J.E. (1994) Expression cloning of a gibberellin 20-oxidase, a multifunctional enzyme involved in gibberellin biosynthesis. *Proceedings of the National Academy of Sciences USA* 91, 8552-8556.

Lester, D.R., Ross J.J., Ait-Ali, T., Martin, D.N. and Reid, J.B. (1996) Pea gibberellin 20-oxidase. *Plant Gene Register* 96-050.

MacMillan, J. (1997) Biosynthesis of the gibberellin plant hormones. *Natural Product Reports* 14, 221-243.

Martin, D.N., Proebsting, W.M., Parks, T.D., Dougherty, W.G., Lange, T., Lewis, M.J., Gaskin, P. and Hedden, P. (1996) Feed-back regulation of gibberellin biosynthesis and gene expression in *Pisum sativum* L. *Planta* 200, 159-166.

Phillips, A.L., Ward, D.A,. Uknes, S., Appleford, N.E.J., Lange, T., Huttly, A.K., Gaskin, P., Graebe, J.E. and Hedden, P. (1995) Isolation and expression of three gibberellin 20-oxidase cDNA clones from *Arabidopsis. Plant Physiology* 108, 1049-1057.

Reed, J.W., Foster, K.R., Morgan, P.W. and Chory, J. (1996) Phytochrome B affects responsiveness to gibberellins in *Arabidopsis. Plant Physiology* 112, 337-342.

Sponsel, V.M., Schmidt, F.W., Porter, S.G., Nakayama, M., Kohlstruk, S. and Estelle, M. (1997) Characterisation of new gibberellin-responsive semidwarf mutants of *Arabidopsis. Plant Physiology* 115, 1009-1020.

Sun, T.P. and Kamiya, Y. (1994) The *Arabidopsis GA1* locus encodes the cyclase *ent*-kaurene synthetase A of gibberellin biosynthesis. *The Plant Cell* 6, 1509-1518.

Talon, M., Koornneef, M. and Zeevaart, J.A.D. (1990) Endogenous gibberellins in *Arabidopsis thaliana* and possible steps blocked in the biosynthetic pathways of the semidwarf *ga4* and *ga5* mutants. *Proceedings of the National Academy of Sciences USA* 87, 7983-7987.

Toyomasu, T., Kawaide, H., Sekimoto, H., von Numers, C., Phillips. A.L., Hedden, P. and Kamiya, Y. (1997) Cloning and characterization of a cDNA encoding gibberellin 20-oxidase from rice (*Oryza sativa* L.) seedlings. *Physiologia Plantarum* 99, 111-118.

Wu, K., Gage, D.A. and Zeevaart, J.A.D. (1996) Molecular cloning and photoperiod-regulated expression of gibberellin 20-oxidase from the long-day plant spinach. *Plant Physiology* 110, 547-554.

Xu, Y.-L. and Zeevaart, J.A.D. (1997) Negative feedback loop of *GA5* expression constitutes part of the mechanism of stem height regulation in *Arabidopsis thaliana*. *Plant Physiology* 114 (suppl.), 163.

Xu, Y.-L., Li, L., Wu, K., Peeters, A.J.M., Gage, D. and Zeevaart, J.A.D. (1995) The *GA5* locus of *Arabidopsis thaliana* encodes a multifunctional gibberellin 20-oxidase: molecular cloning and functional expression. *Proceedings of the National Academy of Sciences USA* 92, 6640-6644.

Xu, Y.-L., Gage, D.A. and Zeevaart, J.A.D. (1997) Gibberellins and stem growth in *Arabidopsis thaliana*. Effects of photoperiod on expression of the *GA4* and *GA5* loci. *Plant Physiology* 114, 1471-1476.

Yamaguchi, S., Saito, T., Abe, H., Yamane, H., Murofushi, N. and Kamiya, Y. (1996) Molecular cloning and characterization of a cDNA encoding the gibberellin biosynthetic enzyme *ent*-kaurene synthase B from pumpkin (*Cucurbita maxima* L.). *The Plant Journal* 10, 203-213.

Index

Abscisic acid (ABA) 143-156, 160
Abscission 63, 68, 146, 176
ACC see 1-Aminocyclopropane-
    1-carboxylate
Aconitase, in banana 42
Acyl carrier protein (ACP), in
    strawberry 57-58
AFLP see Amplified fragment length
    polymorphism
Agritope Company 36
Agrobacterium spp. 31
    A. tumefaciens 97, 210, 213
Amino acids 56, 66, 67, 70-71, 104, 106,
    107, 108
1-Aminocyclopropane-1-carboxylate
    (ACC) 42, 64, 65
Amplified fragment length polymorphism
    (AFLP) 4
Angiosperms 175, 182
Antheraxanthin 143, 144
Anthocyanins 55, 162
Antioxidants 18, 55
Apple 56, 64, 111, 158, 159, 205
Arabidopsis spp. 3, 4, 9, 10, 36, 44, 46,
    65-66, 66-68, 68-72, 126, 136,
    137, 143, 144, 146-151, 182, 183,
    208, 210-215
    A. thaliana 96, 97, 101, 206, 209,
        210-213
    A. thaliana cv. Columbia 104, 105,
        106, 107, 210

Asparagine 106
Aspartic acid 66, 106
Auxin 95, 96, 97, 98, 99, 100, 101, 170
Avocado 56, 65

ß-Amylase, in banana 43-44
ß-1,3-Glucanase, in banana 45
Banana (*Musa* spp.) 41-49, 64
    clone pBANUU129 45
    physiology of 41
    pulp 42-46
Barley 149-150, 166, 180
Bean (*Phaseolus vulgaris*) 208, 209
BER see Blossom-end rot
BERRYSIM software package 119, 120
Beta vulgaris see Sugar beet
Biosynthesis
    of ABA 143-145
    ethylene 1, 3, 32, 35-36, 64-65
    gibberellin 205-217
Biotechnology 151-153
Blossom-end rot (BER), tomato 19, 22-26
Blotchy ripening, tomato 19, 22
Brassic ein mutant 139
Brassica rapa 126
Bryophyllum 133

C:N ratio 106
Calcium 22, 23-24, 25-26
Calcium oxalate 26

Calgene Company 33
Callus formation 88-90, 99, 100, 101
Campanula isophylla 200, 201, 202
CaMV *see* Cauliflower mosaic virus
Canopy photosynthesis, tomato 20, 21
Carbohydrates 43-44, 76, 79, 93,
    104-105, 107, 162
Carnation 65
Carotene 18, 34, 144
Cauliflower mosaic virus (CaMV) 34, 35,
    69, 72, 134, 183, 186, 210
cDNA cloning 5, 11, 31, 42, 43, 44, 46,
    53, 56, 58, 67, 70, 95, 97, 123
cDNA libraries 124, 126
Cell enlargement, tomato 24, 25-26
Cell wall
    degradation 43, 44-45, 64
    pectinase (PG) 3
    properties 11, 22, 32, 44-45,
        157-158, 163, 168-170
Cellulase, in strawberry 55-56
Cereals 44, 186, 205
Chalcone reductase 54
Chalcone synthase 54, 55
Chamaerion angustifolium 181
Chemical regulation 158-162
'Chemical signalling' 158, 159
Chenopodium album 180, 181
Chilling 113, 114-115, 116, 119, 121,
    122
Chitinase 10, 45-46
Chlormequat *see* Cycocel
Chlorophyll 64, 176, 193, 213
Chromophore synthesis 182
Chromoprotein photoreceptors 177
Chrysanthemum spp. 134
    C. morifolium 193, 194, 195, 196,
        198, 209
Circaea lutetiana 181
Citrus fruits 64
'Climacteric' fruits 1, 3, 32, 41, 56, 63,
    64, 65, 70
Cloning *see* cDNA cloning
CLUSTAL X 209
CO_2 enrichment 17, 20
CO_2 fixation 104
Colour 54-55, 69
Computer control, of growing
    environment 27
Conifers 205

Consumer preferences, tomato 17, 18,
    19, 54, 170
Controlled propagation environment
    (CPE) 78, 80-90
Cotinus coggyria cv. Royal Purple 84,
    85-86, 87, 90, 91, 92, 93
Cotton 146, 169
Cowpea (*Vigna sinensis*) 202
CPE *see* Controlled propagation
    environment
CPS *see ent*-Copalyl diphosphate synthase
Craterostigma plantagineum 147, 150
Crop
    performance 175-190
    of softwood cuttings, nature of 76-
        77
    stands *see* Plant density
Cropping patterns, strawberry 118-119,
    127
Crypomeria japonica cv. Elegans
    Compacta 85, 86, 87, 90
Cryptograms, vascular 175
Cucumber (*Cucumis sativus*) 139, 168,
    195
Cucurbita maxima see Pumpkin
Cuttings
    dark-preconditioning 78-79, 87-88,
        90, 183
    defoliation pretreatment 79-80
    leafy 75-94
    stem necrosis 88
    water balance of 77-78
    see also Propagation, vegetative
Cycocel 186, 191
Cysteine proteinase 59
Cytokinins 104, 105-106, 107, 108, 170

Dark patch, tomato 22
Day-night transition *see* Transition
DDTFR *see* Differential display tomato
    fruit ripening
De-etiolation 183, 184, 185, 186
Defoliation, in strawberry 118
Dehydration 151
Depth underwater *see* Underwater depth
Desiccation 92, 145-146, 146-147, 152,
    153
Differential display tomato fruit ripening
    (DDTFR) 10
Dimerization, in tomato 68, 71, 72
DNA gel-blot hybridization 6, 7, 8, 10

Index

Dormancy 112-116, 133, 145-146, 151
DROP treatments 197-199
Drought 157-158, 160, 161, 162, 166, 167
Dry matter content, tomato 19-20
Dwarfism 214, 215
    in cereals 186
    in pea 205
    in potato 139
    in tobacco 183, 186, 187
    in tomato 35, 147, 183, 200

EKS see ent-Kaurene synthase
Electrical conductivity (EC) treatments, tomato 20, 21
Elongation factor 2 (EF-2) 59
Embryos 145-146, 147, 148, 150-151
ent-Copalyl diphosphate synthase (CPS) 206, 207
ent-Kaurene synthase (EKS) 206, 207
Environmental
    control 12, 90-92, 111-131
    'fingerprints' 84
    requirements 75-94
    response data 118-119
    sensing 90-92
    stress 157-174
ERS gene 3, 4
Ethylene sensitivity 65-66
Ethylene signalling
    in banana 41, 42, 46
    in tomato 1-4, 6-10, 12, 63-74
ETR1 gene 3, 4, 66-68, 68-72
Euphorbia pulcherrima see Poinsettia
Evapo-sensor 92
Evaporative demand, measurement of 91-92
Expansins 56-57, 167, 168-169
Extensins, in banana 45

Flavanone-3-hydroxylase 54, 55
Flavonols, in strawberry 55
Flax 180
Floral transition 103-109
'Florigen' 103
Flower initiation 112-114, 195
Flowering, in strawberry 111-131
Fluorimetry 106
Fog systems 91
Forsythia × *intermedia* cv. Lynwood 87

FR-HIR see 'High irradiance response'
Fragaria
    F. ananassa see Strawberry, cultivated
    F. chiloensis 51
    F. vesca 51, 111, 123, 124-127, 128
    F. vesca semperflorens 123, 124, 125, 126, 127, 128
    F. virginiana 51
    see also Strawberry
Frost hardiness 133
Fructose 58, 59
Fruit defects, in tomato 18, 22, 27
Fruit development/growth 1-15, 20-22, 24, 52, 162-167
Fruit quality 31-39
    strawberry 51-61
    tomato 1-15, 17-29
Fruit rot 158
Fuchsia (*Fuchsia* × *hybrida*) 195, 201
Fungi 143, 206

GA see Gibberellin
GAD see Glutamate decarboxylase
GCRI/Bewley Lecture 31-39
GenBank 10
Gene expression 33, 35, 36, 152
    ABA-induced 147-148
    in banana 41-49
    in strawberry 52-53
Gene tagging 95, 96
Genetic manipulation 1-15, 31-39, 119-127
    of cell wall properties 168-170
    of shade avoidance 183-186
    in strawberry 111-131
Genetic mapping 4-6
Geranium (*Pelargonium* × *hortorum*) 65, 195, 199
Geranylgeranyl diphosphate 206
Germination 63, 145-146, 176, 205
Gibberellin (GA) 35, 103, 186, 205-217
    biosynthetic pathway 139-140
    metabolism 199-201
Glucanase, in banana 46
Glucose 58, 59
Glutamate decarboxylase (GAD) 46
Glutamic acid 106
Glutamine 106
Glycolysis, in strawberry 60
Goldspot, tomato 22-23, 26

Gradient CPE (G-CPE) chamber 81–85, 87, 88, 89, 90, 91
Grape 158, 161, 162, 170
'Green Revolution' 186
Growth, manipulation of 157–174

Hales, Stephen 175–176
Harvest index 182, 188
Hexoses, in strawberry 59
'High irradiance response' (HIR) 184, 185, 186
High-performance liquid chromatography (HPLC) 104, 106
Highwire canopy stands 17
Histidine kinase, in tomato 66, 67, 68, 70, 71
Histochemical staining 101
Hormonal regulation 12, 52
Hormone carriers 160
HPLC *see* High-performance liquid chromatography
Humidification 76–77, 91
Hydroponic culture 17, 23, 26

IAA *see* Indoleacetic acid
IBA *see* Indolebutyric acid
Impatiens (*Impatiens wallerana*) 195, 199
Indoleacetic acid (IAA) 97, 98
Indolebutyric acid (IBA) 76, 97, 99
Inter-simple sequence repeat–polymerase chain reaction (ISSR-PCR) 124, 126, 127
Internode length 194, 195, 196, 197
Invertase 44, 58–59
Ions, inorganic 104, 106–107
Irradiance 91, 180, 196
  see also 'High irradiance response'
ISSR-PCR *see* Inter-simple sequence repeat–polymerase chain reaction

Kaempferol, in strawberry 55
Kanamycin 210

Laboratory *see* Controlled propagation environment laboratory
Leaf to air vapour pressure difference (LAVPD) 77, 78, 86, 92

Leaves, growth of 150–151, 162–167
Light signal 5, 12, 177–178
Lily (*Lilium longiflorum*) 193, 195, 197, 199
Lodging 186
Lolium spp. 139, 166
  L. temulentum cv. Ceres 104, 105, 107
Lotus corniculatus 107
Lyase, pectate, in banana 44–45
Lycopene 18, 19, 22, 34
Lycopersicon
  L. cheesmannii 7, 9
  L. esculentum cv. VFN8 9
  L. pimpinellifolium see Tomato, wild
  L. pinnellii 7, 68
  see also Tomato, cultivated
Lysine 152, 153

Maize 44, 147, 150, 152, 153, 164, 167, 180
MAPK *see* Mitogen-activated protein kinase
Marah macrocarpus 209
Melon, cantaloupe 36
Mercurialis perennis 181
Micropressure probe 162, 163
Mildew 158
Mist systems 90–91
Misting equipment, automatic 78
Mitogen-activated protein kinase (MAPK) 149
Molecular
  biology 2–3, 31–39, 95–102
  mapping 124–127
  markers 124, 126
Monsanto Company 36
mRNAs 32
Mung bean 44
Musa spp. *see* Banana
Mutants 95–102, 182

1-Naphthaleneacetic acid (NNA) 52, 98
Nasturtium 168
Neoxanthin 143, 144
  cleavage enzyme (NCE) 160
Netting, tomato 22
Never-ripe (*Nr*) *see* Tomato

Nicotiana
  N. affinis 209
  N. plumbaginifolia 143, 149
  N. sylvestris 138
  N. tabacum cv. Xanthii 96, 97, 98
Night-day transition *see* Transition
Northern blot analysis 54, 57, 59, 160, 170, 212
Nr gene *see* Tomato, *Never-ripe*
Nutrients, tomato 25

Oats 183, 186, 188
Osmotic stress 23, 25, 26

Paclobutrazol 201
PAL *see* Phenylalanine ammonia lyase
Pan evaporimeter 92
PAR *see* Photosynthetically active radiation
Partial root-zone drying (PRD) 161-162
Pea (*Pisum sativum*) 138, 139, 205, 208
Peach 115
Pelargonium × *hortorum see* Geranium
Penman evapotranspiration estimate 92
Pepper 56
Perennial cycle, in strawberry 112, 117-118
Peroxidase 165-167, 168, 169-170
Petunia (*Petunia* × *hybrida*) 72, 195, 197, 199
PG *see* Cell wall pectinase; Polygalacturonase
pH 160, 161
Phaseolus vulgaris see Bean
Phenolics 41, 52, 55, 162
Phenoxyacetic acid (POA) 52, 53
Phenylalanine ammonia lyase (PAL) 52, 54
Phenylpropanoid metabolism, in strawberry 54-55
Phloem exudate (sap) 20, 103, 104, 107
Phosphorylation 149
Photochemical properties 177
Photoperiodism 103, 104, 112, 118, 133-142, 183
Photosynthesis 80, 83, 175, 179, 181, 193
Photosynthetic photon flux density (PPFD) 84, 85, 86, 88

Photosynthetically active radiation (PAR) 76
PHY transgenes 188
Physiological analysis 103-109
Phytochrome
  A 183-188
  B 134-137, 139-140, 141, 182, 184, 185
  genes 175-190
Phytochromes, absorption spectra 177-178
Phytochromes D, E 183, 185
Pisum sativum see Pea
Plants
  density 21, 122, 175-177, 186, 187
  morphology 205-217
  ornamental 191, 193, 205
  reproduction 133-142
  transgenic 2, 11, 33, 57, 59, 72, 107, 135-137, 148, 151, 152, 169, 170, 185, 187, 188, 213
POA *see* Phenoxyacetic acid
Poinsettia (*Euphorbia pulcherrima*) 134, 186, 191, 198
Pollination 72
Polyamines 104, 106, 107
Polyethylene glycol (PEG) 146
Polygalacturonase (PG) 33-35
Polysaccharides, in strawberry 52, 57
Potassium 22
Potato 44, 107, 134, 135, 138, 140, 141
Potentilla fruticosa 124
PPFD *see* Photosynthetic photon flux density
Promoters 119, 149-150
Propagation environment, controlled 80-84
Proteinase *see* Cysteine proteinase
Proteins 32, 42, 56, 58, 59, 60, 66, 70
  see also Acyl carrier protein
Proximity sensors 180
Pruning 158
Psychrometer 157
Pumpkin (*Cucurbita maxima*) 208, 209, 210, 212, 214, 215
Pyruvate decarboxylase (PDC) 59-60

Qualitative trait locus (QTL) 2
Quality *see* Fruit, quality; Vegetable quality
Quercetin 5, 6, 55

rab17 gene 150-151
rab28 gene 150
rac mutant 98-99, 99-101
Radish 180
Ragweed, short 44
Random amplified polymorphic DNA (RAPD) 4
RAPD *see* Random amplified polymorphic DNA
Raspberry 56
Refractometry 104
Regulated deficit irrigation (RDI) 158
Relative growth rate (RGR) 164
Restriction fragment length polymorphism (RFLP) 4, 6, 9
Reverse transcriptase-polymerase chain reaction (RT-PCR) 169
RFLP *see* Restriction fragment length polymorphism
Rice 169, 183
Ripening 63
  in banana 41-49
  in strawberry 51, 52, 53, 58, 59-60
  in tomato 1, 2, 3, 10-11, 31-34, 64, 65, 69
Root drying, partial 161-162
Root formation 95-102
Rooting
  adventitious 75, 95, 96, 97, 98-99, 100, 101, 102
  lateral 95, 96
  potential 75-94
rty-3 mutant 97-98
Runner production, in strawberry 111, 112, 115-117, 119-127

Salinity 23, 24, 151, 153
Salvia (*Salvia splendens*) 195, 197, 199
SAM *see* Shoot apical meristem
Sap *see* Phloem exudate
Scan, spectroradiometric 177
Senecio vulgaris 181
Senescence 31, 35, 46, 59, 63, 139, 176, 180
Serine 106
Shade avoidance 175, 176, 178, 180-188
Shade tolerance 175, 180
Shoot apical meristem (SAM) 103, 107
Signal pathways, ABA 148-152
Signals, inductive/inhibitory 137-138, 140

Sinapis alba 104, 105, 106, 107, 108
Softening 19, 22, 64
Soil drying 157-164, 167, 170
Solanum
  S. demissum 134
  S. dulcamara 206, 209, 213-214
  S. tuberosum ssp. *andigena* 134, 138, 139
Sorghum 136
  ma_3^R mutant 139
Southern blot analysis 68
Spinach (*Spinacea oleracea*) 139, 208
Starch 41, 43-44, 105, 107
Stem extension 191-203
Strawberry (*Fragaria* spp.) 45, 51-61
  aroma of 58
  biochemistry of 52, 60
  cultivated (*F. ananassa*) 51, 123, 127
  defoliation 118
  cv. Elsanta 112-117, 119, 120, 121, 122
  flavanols 55
  flowering 113-131
  glycolysis 60
  hexoses 59
  runner production 111, 112, 115-117, 119-127
  texture 55-56, 57
  transferases 54-55
  see also Fragaria
Stress tolerance 151, 152
Sucrose 5, 58, 59, 104, 105, 106, 107, 108
Sugar beet (*Beta vulgaris*) 180, 209
Sunflower 161, 180
Sweden 199
Syringa vulgaris cv. Madame Lemoine 79, 80, 88, 89, 90, 93

TCTR1 gene 6-8
Temperature control 83-84
Temperature regimes, tomato 17, 26, 166
Teucrium scorodonia 181
Texture, in strawberry 55-56, 57
Thaumatin, in banana 46
'Thermoperiodicity' 193
Thin-layer chromatography (TLC) 106
Thiol proteinase *see* Cysteine proteinase
'Tissue pressure' (TP) 163

Tobacco 44, 45, 100, 101, 107, 138, 141, 152, 169, 180, 183, 186, 188
   Hicks *MM* 136, 137
   Maryland Mammoth 134, 135, 136, 138
   see also Nicotiana
Tomato 44, 56, 59, 146, 161-168, 180, 183
   cultivated (*Lycopersicon esculentum*) 1-15, 17-29, 63-74, 191-197, 201
   cv. Moneymaker 192, 200
   Daniella 22
   Epi mutant 4, 8-10
   Flavr Savr 34
   glasshouse 17, 18
   Never-ripe (*Nr*) 3, 4, 11, 36, 65, 68-70, 72
   nor (non-ripening) mutant 3, 11
   rin (ripening-inhibitor) mutant 3, 11
   r^y 34
   wild (*Lycopersicon pimpinellifolium*) 2
   yellowflesh 34-35
   see also Lycopersicon
Transferases, in strawberry 54-55
Transition, day-night/night-day 178-179, 191-197, 200
Transpiration 23-24, 76, 77
Truss pruning, tomato 24
Tuberization 133-135, 137, 138, 140, 141, 170
Tyrosine aminotransferase 98

Ubiquitin pathway 186

Underwater depth 178-179
Urtica dioica 181

Vegetable quality 31-39
Vegetative growth, in strawberry 112, 114-115, 118-119
Verbena spp. 199
Vernalization 133, 205
Vigna sinensis see Cowpea
Violaxanthin 143, 144
Vitamins, in tomato 18, 19

Water balance 77-78
Water stress 25, 76, 77, 78, 81, 88, 90-91, 92, 143-156
Weigela florida cv. Variegata 87
'Wet fog' 77
Wheat 180

Xanthium strumarium 104, 105, 106, 107
Xanthoxin 143, 144
XET *see* Xyloglucan endoglycosylase
Xylem 160, 161, 162, 167
Xyloglucan endoglycosylase (XET) 164-165, 168

Yeast artificial chromosome (YAC) colonies 5, 124

Zeaxanthin 143, 144, 148
Zeaxanthin epoxidase (ZE) 160